Springer Tracts in Modern Physics 88

Springer Tracts in Modern Physics

66* **Quantum Statistics in Optics and Solid-State Physics**
With contributions by R. Graham, F. Haake

67* **Conformal Algebra in Space-Time and Operator Product Expansion**
By S. Ferrara, R. Gatto, A. F. Grillo

68* **Solid-State Physics** With contributions by D. Bäuerle, J. Behringer, D. Schmid

69* **Astrophysics** With contributions by G. Börner, J. Stewart, M. Walker

70* **Quantum Statistical Theories of Spontaneous Emission and their Relation to Other Approaches** By G. S. Agarwal

71 **Nuclear Physics** With contributions by J. S. Levinger, P. Singer, H. Überall

72 **Van der Waals Attraction:** Theory of Van der Waals Attraction By D. Langbein

73 **Excitons at High Density** Edited by H. Haken, S. Nikitine. With contributions by V. S. Bagaev, J. Biellmann, A. Bivas, J. Goll, M. Grosmann, J. B. Grun, H. Haken, E. Hanamura, R. Levy, H. Mahr, S. Nikitine, B. V. Novikov, E. I. Rashba, T. M. Rice, A. A. Rogachev, A. Schenzle, K. L. Shaklee

74 **Solid-State Physics** With contributions by G. Bauer, G. Borstel, H. J. Falge, A. Otto

75 **Light Scattering by Phonon-Polaritons** By R. Claus, L. Merten, J. Brandmüller

76 **Irreversible Properties of Type II Superconductors** By. H. Ullmaier

77 **Surface Physics** With contributions by K. Müller, P. Wißmann

78 **Solid-State Physics** With contributions by R. Dornhaus, G. Nimtz, W. Richter

79 **Elementary Particle Physics** With contributions by E. Paul, H. Rollnick, P. Stichel

80* **Neutron Physics** With contributions by L. Koester, A. Steyerl

81 **Point Defects in Metals I:** Introduction to the Theory 2nd Printing
By G. Leibfried, N. Breuer

82 **Electronic Structure of Noble Metals, and Polariton-Mediated Light Scattering**
With contributions by B. Bendow, B. Lengeler

83 **Electroproduction at Low Energy and Hadron Form Factors**
By E. Amaldi, S. P. Fubini, G. Furlan

84 **Collective Ion Acceleration** With contributions by C. L. Olson, U. Schumacher

85 **Solid Surface Physics** With contributions by J. Hölzl, F. K. Schulte, H. Wagner

86 **Electron-Positron Interactions** By B. H. Wiik, G. Wolf

87 **Point Defects in Metals II:** Dynamical Properties and Diffusion Controlled Reactions
With contributions by P. H. Dederichs, K. Schroeder, R. Zeller

88 **Excitation of Plasmons and Interband Transitions by Electrons**
By H. Raether

* denotes a volume which contains a Classified Index starting from Volume 36.

Heinz Raether

Excitation of Plasmons and Interband Transitions by Electrons

With 121 Figures

Springer-Verlag Berlin Heidelberg GmbH 1980

Professor Dr. Heinz Raether

Institut für Angewandte Physik, Universität Hamburg,
Jungiusstraße 11, D-2000 Hamburg 36, Fed. Rep. of Germany

Manuscripts for publication should be addressed to:

Gerhard Höhler

Institut für Theoretische Kernphysik der Universität Karlsruhe
Postfach 6380, D-7500 Karlsruhe 1, Fed. Rep. of Germany

Proofs and all correspondence concerning papers in the process of publication should be addressed to:

Ernst A. Niekisch

Haubourdinstrasse 6, D-5170 Jülich 1, Fed. Rep. of Germany

ISBN 978-3-662-15796-1 ISBN 978-3-540-34716-3 (eBook)
DOI 10.1007/978-3-540-34716-3

Library of Congress Cataloging in Publication Data. Raether, Heinz. Excitation of plasmons and interband transitions by electrons. (Springer tracts in modern physics; v. 88) Bibliography: p. Includes index. 1. Plasmons (Physics) 2. Electronic excitation.
I. Title. II. Series. QCI. S797 vol. 88 [QCI76.8.P55] 539'.08s [530.4'1] 79-22131

Originally published by Springer-Verlag Berlin Heidelberg New York in 1980
Softcover reprint of the hardcover 1st edition 1980

2153/3130 — 5 4 3 2 1 0

Preface

This book treats the physics of Electron Loss Spectroscopy (ELS) with electrons of different energies. Its emphasis is on the collective excitations or plasmons in the bulk as well as on the surface, including interaction with the interband transitions. A discussion of the experimental and theoretical aspects helps to define the current state of the art.

In addition to the general physics, data on the plasmons (energy, halfwidth, dispersion) obtained by different methods of observation (transmission and reflection with fast and slow electrons) have been collected as completely as possible. Mr. R. Manzke was helpful in gathering these figures. Related topics such as coupling of plasmons with light, Cerenkov radiation and waveguides, plasmons in electron gases of one and two dimensions together with some applications (microanalysis) are briefly described to round off the representation and to demonstrate the usefulness of the plasmon concept. I would like to thank Dr. I. Pockrand for reading Chap. 10 and for suggestions.

I hope that this volume which replaces an earlier article by the author (Springer Tracts in Modern Physics, Vol. 38, 1965) will be a helpful guide for further work and progress in ELS. At the same time it is a reminder to me of the exciting times spent with my co-workers in this field.

Hamburg, December 1979 H. Raether

Contents

Introduction

Electron loss spectroscopy (ELS) gives an important insight into the variety of excitations of a solid. The information is obtained by measuring the energy losses ΔE which the electrons, fast or slow, suffer by transmitting a thin film of the solid or by being reflected at its boundary. The transmission is mainly used in experiments with fast electrons of some 10 keV and more, whereas the reflection method is well adapted for low energy electrons of about 100 eV and less which cannot penetrate thin films in general.

If the electrons pass the crystal or the condensed material they interact with the free and the bound electrons of the solid. These "electronic" excitations cover a large energy range from about 1 to several 10^2 eV. Further we can observe the excitation of vibrations of heavy particles (ions), e.g., the surface phonons on surfaces of alkali halides which have an energy of about 1/100 eV. Energy losses are also produced by coupling of the incoming electrons with the electromagnetic field so that photons are created if the velocity of the electrons v_{el} is higher than the velocity of light inside the solid with the refractive index n:c/n or $v_{el} > c/n$ (Cerenkov radiation). This condition is fulfilled in the region of visible light in a number of semiconductors as Ge, Si, and III-V compounds and leads to energy losses of some eV.

It is characteristic for the angular distribution of the inelastically scattered electrons that they are concentrated in an aperture of about $\theta_{\Delta E} = \Delta E/2E_0$ (ΔE energy loss, E_0 kinetic energy of the incoming electrons). In contrast the elastically scattered electrons are spread out into an angle of $\theta \sim 1/k_{el} \cdot a$ which is about 10^2 larger than $\theta_{\Delta E}$ (k_{el} wave vector of the incoming electrons, a radius of the scattering atom about $0.5/Z^{1/4}$ Å.

The main concern of this article consists in describing *electronic* excitation that means the energy (ΔE) and momentum ($\hbar q$) is transferred to the electrons of the solid; this process can be roughly described as follows:

1) In the case of a metal the incoming electrons interact with the more or less free valence electrons which density amounts to 10^{22} - 10^{23} cm^{-3} (electron plasma).

a) For low values of q transfer $q < q_c$ (cutoff vector) or low values of the scattering angle $\theta < \theta_c$, $\theta_c = q_c/k_{el}$, collective oscillations of the electron plasma or "plasmons" can be excited. These oscillations run as longitudinal charge density fluctuations through the volume of the crystal (volume plasmons) and along its surface (surface plasmons). Its energy $\hbar\omega$, ω frequency of the oscillation, is of the order of 10 eV and depends on the density of the loosely bound electrons. We understand as loosely bound electrons those for which $\hbar\omega_p$, ω_p plasma frequency, is large compared with their binding energy, e.g., in metals the electrons in the conduction band, in semiconductors the electrons in the valence band. The frequency of the volume waves is higher than that of the surface waves; the dispersion relations of both are rather different and their excitation probability too.

b) If the momentum transfer $\hbar q$ exceeds the critical value $\hbar q_c$, q_c is given approximately by ω_p/v_F, v_F Fermi velocity, the incoming electrons excite no more collective oscillations, but transfer ΔE and $\hbar q$ to single electrons of the electron plasma. If a plasmon with a wave vector $q \geq q_c$ is excited, it decays into an electron-hole pair, so that the final result is a single particle excitation. This situation can be described by using $V(r) \sim (1/r)\exp(-k_s r)$, the interaction potential of the incoming electron and the scattering electron; here r means the distance between the two electrons and $(k_s)^{-1}$ the screening length or the radius of the polarization cloud. This Thomas-Fermi length has a value of about 1 Å. If $q > q_c$ the collision parameter becomes comparable to and smaller than $(k_s)^{-1}$ so that the incoming electron collides with a nearly unscreened plasma electron and makes a single electron excitation. The transferred energy values vary from zero to high energies.

2) In large gap insulators the conditions are not favourable for the excitation of plasmons. Here we observe interband transitions in the energy region of the valence band ($\sim 10^1$ eV), excitons included, and of deeper levels, e.g., K, L ... edges (some 10^2 eV). The deeper the shells the more the excitation can be regarded as that of a single particle; in general the probability of excitation can therefore be calculated assuming an interaction of the incoming electrons with isolated atoms or ions.

The position of these edges can be used to identify the chemical nature of the irradiated atom of the solid. The height of the excitation signal will be a measure of the irradiated quantity so that a chemical microanalysis is possible. More detailed study of this signal has revealed a fine structure on its high-energy side (EXAFS) which indicates that the excited atom is not isolated, but in the neighbourhood of other atoms; their coordination can be derived from the fine structure of the loss spectrum.

In the region of low-energy losses of about 10 eV (plasmon and interband transitions) the results of the experiments are described by the dielectric theory. One has to keep in mind that real plasmons move in a periodic potential and not in a free electron gas as the ideal plasmons, so that the dielectric function used for the calculation of energy, lifetime, q dependence and excitation probability, etc, has to consider the band structure. The ELS with fast electrons has given detailed information of the properties of the real plasmons and contributed considerably to develop the physics of plasmons.

In the following a more or less complete survey of the results of the ELS is given. At first the collective excitations in the volume (volume plasmons) and their various properties detected up to now are discussed. Further the essential results of the wave vector dependence of the loss energy $\Delta E = \hbar\omega(q)$ and its halfwidth are summarized. The excitations of bound electrons which lose their collective character with increasing value of ΔE are treated together with some applications. The surface plasmons are described in a separate chapter. A short survey of the experimental methods finishes the article.

1. Volume Plasmons

1.1 General Properties

The following lines represent a survey of the phenomena called the collective excitations of the dense electron gas of a solid in the volume /1.1/. Such an electron gas or electron plasma is nearly realized, e.g., by the valence electrons of a metal ($n_0 \sim 10^{22}$ cm^{-3}) or by the electrons of a semiconductor whose conduction band contains electrons of a donator (up to $\sim 10^{19}$ cm^{-3}). This excitation consists of a coherent motion of a very high number of electrons in contrast to those excitations in which the external perturbation acts on a single electron of the whole assembly of electrons; it leads to small density fluctuations around the normal bulk density n of the electron plasma. The energy quanta of these volume plasma oscillations are called volume plasmons. The positive ions neutralise the mean negative charge. The density fluctuations Δn of n are described by plane longitudinal waves with the amplitude Δn_{max}

$$\Delta n(x,t) = n(x,t) - n = \Delta n_{max} \exp i(\omega t - qx) \tag{1.1}$$

here ω means the frequency and $q = 2\pi/\lambda$ the wave vector. A high number of electrons participates at this oscillation, so that it is justified to speak of a collective mode.

In a first approach the electrons of the plasma are treated as free electrons neglecting the periodicity of the potential of the lattice ions. We shall see that the band structure cannot be neglected since it determines important properties of the plasmons as, e.g., their lifetime. These calculations are performed in general in the random phase approximation (RPA) which neglects electron-electron correlation (Coulomb interaction - and spin exchange). It means that the density deficit produced around an electron by these interactions is not taken into account. Experimental results however suggest considering these correlations, (compare dispersion at small and large q vectors, Chaps. 6 and 7). If the electron plasma has boundaries further modes, the surface polaritons, exist. For large wave vectors they are called surface plasmons since their electromagnetic fields which originate in charges oscillating in the boundary are concentrated at the surface. These surface plasmons which exist as radiative and as nonradiative ones are treated in Chap. 10; see also /1.2/.

1.2 The Eigenfrequency of Plasma Oscillations

The bulk plasmons described by (1.1) have an eigenfrequency which can be calculated using Maxwell's equations. The longitudinal character of these oscillations needs rot H = 0 (H magnetic field). Then we have with $\dot{D}$ (displacement current) = $\tilde{\varepsilon}_0 \dot{E}$, ($\tilde{\varepsilon}_0 = (4\pi\ 9\cdot 10^9)^{-1}$ Coul $(\text{volt}\cdot\text{m})^{-1}$) in vacuum and with the convection current j produced by the movement of charges

$$\text{rot } H = \dot{D} + j = i\omega\varepsilon(\omega)\, E = 0. \tag{1.2}$$

Since the amplitude E of the charge fluctuations shall be finite in the absence of external fields the condition

$$\varepsilon(\omega) = 0 \tag{1.3}$$

gives the eigenfrequency.

The electric field E of these longitudinal oscillations lies parallel to $\vec{q}$. There is no magnetic field since the H field of j and that of dD/dt compensate each other at every moment.

To obtain the eigenfrequency one needs, as we see, the dielectric function. In the simplest case we assume a free electron plasma with a relaxation time τ and restrict to long wavelenghts ($q \sim 0$). Then we have the well known dielectric function

$$\varepsilon = \varepsilon' + i\varepsilon'' = 1 - \frac{\omega_p^2}{\omega^2} \cdot \frac{1}{1 - 1/i\omega\tau} \tag{1.4}$$

with

$$\varepsilon' = 1 - \frac{\omega_p^2}{\omega^2} \frac{1}{1 + (1/\omega\tau)^2} \tag{1.5}$$

$$\varepsilon'' = \frac{1}{\omega\tau} \frac{\omega_p^2}{\omega^2} \frac{1}{1 + (1/\omega\tau)^2} \tag{1.6}$$

We obtain as solution $\omega = \omega' + i\omega''$ of (1.3)

$$\omega' = \omega_p \left(1 - 1/(2\omega\tau)^2\right)^{1/2} \tag{1.7a}$$

$$\omega'' = -\ 1/2\tau \tag{1.7b}$$

or

$$\omega' = \omega_p \qquad \omega_p\tau >> 1$$

with

$$\omega_p = \frac{ne^2}{m\tilde{\varepsilon}_0} \tag{1.8}$$

ω'' describes the damping of the amplitude of the oscillation: $\exp(-\omega''t) = \exp(-t/2\tau)$. It displaces the plasma frequency to slightly smaller values as (1.7) demonstrates.

The electrons being screened against each other by a (positive) polarisation cloud[1] move nearly independently. Coherent movements are strongly damped as long as the frequency of their motion is below ω_p. At $\omega = \omega_p$, the eigenfrequency of longitudinal oscillations in the electron gas, the long-range Coulomb forces become effective and undamped waves travel through the plasma. (Equation (1.18) shows that q takes real values for $\omega \geq \omega_p$.)

1.3 The Dispersion Relation. The Loss Function

In general the eigenfrequency changes its value with the wavelength. This dispersion relation $\omega(q)$ can be obtained, if we know the q dependence of the dielectric function $\varepsilon(\omega,q)$ or its spatial dispersion

$$\varepsilon(\omega,q) = 0. \tag{1.9}$$

To excite the eigenfrequency of this system an exterior field has to act on it. As we shall see later the excitation of the plasma oscillation is determined by the loss function

$$\mathrm{Im}(-1/\varepsilon) = \frac{\varepsilon''}{|\varepsilon'|^2 + |\varepsilon''|^2} \tag{1.10}$$

the negative imaginary part of the reciprocal value of $\varepsilon(\omega,q)$. Here we calculate this quantity using the dielectric function of a free electron gas (1.4) and obtain

---

[1] The diameter of the polarisation cloud is given by $(k_s)^{-1} = 0.3\ (r_s/a_0)^{1/2}$ Å with $a_0 = 0.5$ Å and $r_s = \left(\frac{3}{4\pi n}\right)^{1/3}$.

$$\mathrm{Im}(-1/\varepsilon) = \frac{\omega_p^2\ \omega/\tau}{(\omega^2 - \omega_p^2)^2 + (\omega/\tau)^2} \tag{1.11}$$

This Lorentz-like function has a maximum for small $1/\tau$ at ω_p, (1.8), which amounts to

$$\mathrm{Im}(-1/\varepsilon)_{max} = 1/\varepsilon''(\omega_p) = \omega_p\tau. \tag{1.12}$$

Its halfwidth is given by

$$\Delta E_{1/2} = \hbar/\tau \tag{1.13}$$

or

$$(\Delta E_{1/2}/\Delta E)_{max} = 1/\omega_p\tau = \varepsilon''(\omega_p) \tag{1.14}$$

The loss probability reaches this maximum, see Fig. 1.1, where - at small $\varepsilon''(\omega)$ - the denominator of (1.11) goes through a minimum. In contrast the absorption of light which is determined by $\varepsilon''(\omega)$ shows no structure at $\varepsilon' \sim 0$; this transverse excitation displaces the electrons perpendicular to the direction of propagation without changing the electron density.

The first calculation of $\varepsilon(\omega,q)$ of a (Fermi) gas has been given by LINDHARD: An external perturbation produces in a free electron gas intraband transitions and changes the electron state

$$E_k = \frac{\hbar^2}{2m} k^2 \text{ into } E_{k+q} = \frac{\hbar^2}{2m}(k+q)^2 \tag{1.15}$$

by transferring the momentum $\hbar q$. Neglecting damping ($\tau \to \infty$) the dielectric function becomes

$$\varepsilon(\omega,q) = 1 - \lim_{\tau\to\infty} \frac{4\pi e^2}{q^2\Omega} \sum_k \frac{f_o(k+q) - f_o(k)}{E_{k+q} - E_k - \hbar\omega + i\hbar/\tau} \tag{1.16}$$

f_o represents the Fermi-Dirac distribution and Ω the volume of the plasma (notice: the density of states is proportional to Ω, so that Ω in the denominator is compensated).

In the model of a free electron gas as it is assumed, e.g., for the calculation of $\varepsilon(\omega,q)$ of (1.16) damping is excluded. The longitudinal mode $\omega = \omega_p$ obtained from $\varepsilon = 0$ represents thus the ideal volume plasmon (a plasmon of "pure breed").

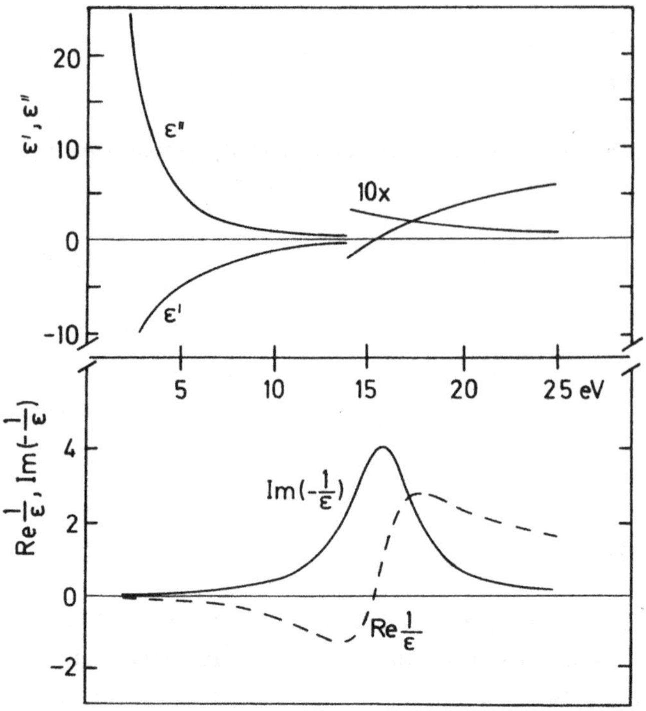

Fig. 1.1. Dielectric function and loss function Im(-1/ε) of a free electron gas for $\hbar\omega_p$ = 15 eV and $\hbar/\tau$ = 4 eV.

Its loss function can be written as

$$\mathrm{Im}(-1/\varepsilon) = \frac{\pi}{2} \cdot \omega_p \cdot \delta(\omega - \omega_p). \tag{1.16a}$$

Integrated it gives the finite value

$$\int_0^\infty \mathrm{Im}(-1/\varepsilon)\omega d\omega = \frac{\pi}{2}\,\omega_p^{\,2} \tag{1.16b}$$

Interaction with the solid is phenomenologically introduced by the relaxation time τ, see (1.4); this has been done in Lindhard's equation too /1.4/. We see later, Chapt. 6, that the width of the loss function is strongly influenced by the band structure of the solid which causes most of the short lifetime of the volume plasmon.

From equation (1.16) one derives for $q \ll q_F$ the dielectric longitudinal function which describes the polarisation under the influence of a longitudinal electric field, as

$$\varepsilon_\ell(\omega,q) = 1 - \frac{\omega_p^2}{\omega^2} \frac{1}{1 - 3/5\,(\vec{q},\, \vec{v}_F/\omega)^2} \tag{1.17}$$

This is a real function since we have set $\tau \to \infty$ in (1.16). From the condition $\varepsilon = 0$ the important equation follows

$$\omega(q)^2 = \omega_p^2 \frac{1}{1 - \frac{3}{5}\,(\vec{q},\, \vec{v}_F/\omega)^2} = \omega_p^2 + \frac{3}{5} v_F^2 q^2 + \dots \tag{1.18}$$

It can be written ($\omega_p^2 > \frac{3}{5} v_F^2 q^2$):

$$\Delta E(q) = \Delta E(o) + (\hbar^2/m)\, \alpha \cdot q^2 \tag{1.19}$$

with

$$\alpha = \frac{3}{5} \cdot \frac{E_F}{\hbar\omega_p} \tag{1.20}$$

and

$$\Delta E(q) = \hbar\omega(q)$$

This is the theoretical dispersion of volume plasmons of a free electron gas valid for $q \ll q_F$. The same relation can be derived for small q applying Boltzmann's equation.

Introducing electron-electron correlations as just mentioned, one has calculated approximations beyond RPA; different results for α have been obtained. The expression derived, e.g., in /1.3/ replaces α by

$$\alpha^* = \alpha \left[1 - \frac{1}{16} \left(\frac{\hbar\omega_p}{E_F} \right)^2 \right] \tag{1.21}$$

which reduces α by about 10 %. For other corrections see Chap. 7, Tab. 7.5.

There is various agreement between the experimental and the values calculated with (1.18), the dispersion relation of a free electron gas. This is more or less to be expected. However, up to now corrections as interband transitions and electron-electron correlations do not yet lead to a really satisfying agreement, see Chap. 7.

If $q \to 0$ the longitudinal ε function of (1.17) and the ε function of (1.4) with $\tau \to \infty$ become equal. Therefore we identify $\varepsilon_\ell(\omega,0)$ with the dielectric function used in optics of visible light since we can neglect momentum transfer in the optical processes in general: if light of energy $\hbar\omega$ and momentum $\hbar q = h/\lambda$ is absorbed by a crystal electron, the electron is transferred to another energy band of the crystal (interband transition) and has increased its momentum by the value h/λ. Comparing h/λ with the elementary cell of the momentum space h/a of a crystal (Brillouin zone) we find that in the visible, $\lambda \sim 6000$ Å,

$$h/\lambda \ll h/a \tag{1.22}$$

assuming a lattice constant of the crystal $a = 4$ Å. Thus no momentum transfer happens practically in light absorption and we speak of a "vertical direct" transition of the crystal electron.

This is different with electrons and hard photons, which produce "nonvertical direct" transitions, so that $\varepsilon(\omega,q)$ has to be regarded and q dependencies become important.

1.4 The Critical Wave Vector

Beyond a theoretical value of the wave vector, the critical wave vector q_c, collective oscillations decay by exciting an electron of the Fermi sea and creating a hole. This limit is calculated as follows: The electron of the plasma may have a momentum (in the q direction) of $\hbar q_i$ before the plasmon transfers the energy ΔE and the momentum $\hbar q$ to it. Applying conservation of energy we obtain

$$\hbar\omega(q) = \frac{\hbar^2}{2m}(\vec{q} + \vec{q}_i)^2 - \frac{\hbar^2}{2m}(\vec{q}_i)^2 = \frac{\hbar^2}{2m}(q^2 + 2\,\vec{q}\vec{q}_i) \tag{1.23}$$

The cutoff vector q_c given by the smallest value of q in (1.23) for $\vec{q} \parallel \vec{q}_i$, is obtained at the maximum value of $\vec{q}_i$ which is $q_i = q_F$.

In a one-dimensional metal the collective oscillations do not cross the single particle boundary, so that their dispersion relation extends theoretically from $\omega = 0$ to very high values, see Chap. 10.

For a rough approximation we assume $q_c < q_F$ and $\omega(q) = \omega_p$ so that (1.23) gives

$$\omega_p = \frac{\hbar}{m} \cdot q_c \cdot q_F \qquad \text{or} \qquad q_c \sim \omega_p/v_F. \tag{1.24}$$

A better approximate value can be calculated by using for $\omega(q)$ the dispersion relation $\omega(q)$ of (1.18). Equation (1.24) can be interpreted: if the phase velocity of the plasma oscillations ω_p/q equals the particle velocity v_F the collective movement decays into that of one electron-hole pair.

Some values of q_c calculated with (1.24) are collected in Table 1.1. These considerations are resumed in Fig. 1.2: It displays the dispersion relation of the volume plasmon $\omega(q)$, curve (1), and the single particle continuum. The width of this band is given by the two curves (2) and (4): at the low-energy side by $\vec{q} \parallel \vec{q}_F$ in (1.23) which means that the plasmon momentum runs parallel to the momentum of the electron on the Fermi sphere to which the plasmon transfers its momentum and ener-

Table 1.1

substance	Be	Al	Li	Si	
cutoff vector q_c:	1.2	1.3	0.9	1.1	$Å^{-1}$

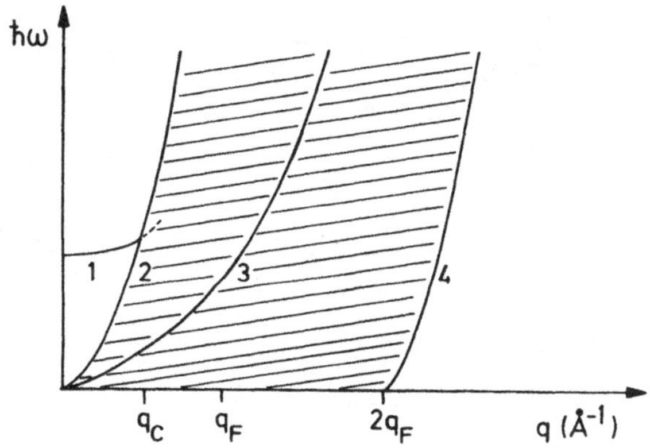

Fig. 1.2. Dispersion relation of the volume plasmon (1) and the band of single electron excitation. (2) and (4) are the limits of this single particle continuum given by $\vec{q}$ parallel to $\vec{q}_F$, curve (2), and $\vec{q}$ antiparallel to $\vec{q}_F$ curve (4), see (1.23). (3) means the "Compton" line without being enlarged by the Doppler effect (energy loss $\hbar\omega$ of an electron scattered by an electron at rest)

gy; the border at the high-energy side is given by $\vec{q}$ and $\vec{q}_F$ antiparallel. The curve (3) of Fig. 1.2 comes out for $q_i = 0$ which means an inelastic collision with an electron at rest ("Compton" line). Collective and single particle excitation are coexistent for $q < q_c$.

Another approach to describe the critical wave vector is the following: The dielectric function of Lindhard for Al with a damping factor $\gamma = 1/\omega_p\tau = 0.033$ is seen in Fig. 1.3; the value of γ is parameter. The shape of the curves ε' and ε'' approaches for small values of q the well-known function at q = 0 reproduced, e.g. in Fig. 1.1. It is remarkable that for small finite q two values $\varepsilon' = 0$ are found; the zero value at smaller energy leads to a broad peak of the single electron excitation, see Fig. 7.6, whereas the higher one leads to the collective excitations (∿ 15 eV). For $q \geq 1.3$ Å^{-1} the ε' function shows no zero value. The q value at which ε' becomes no more zero is identical with the cutoff vector q_c indicating that collective oscillations no longer exist. Further discussion see Sect. 7.2.

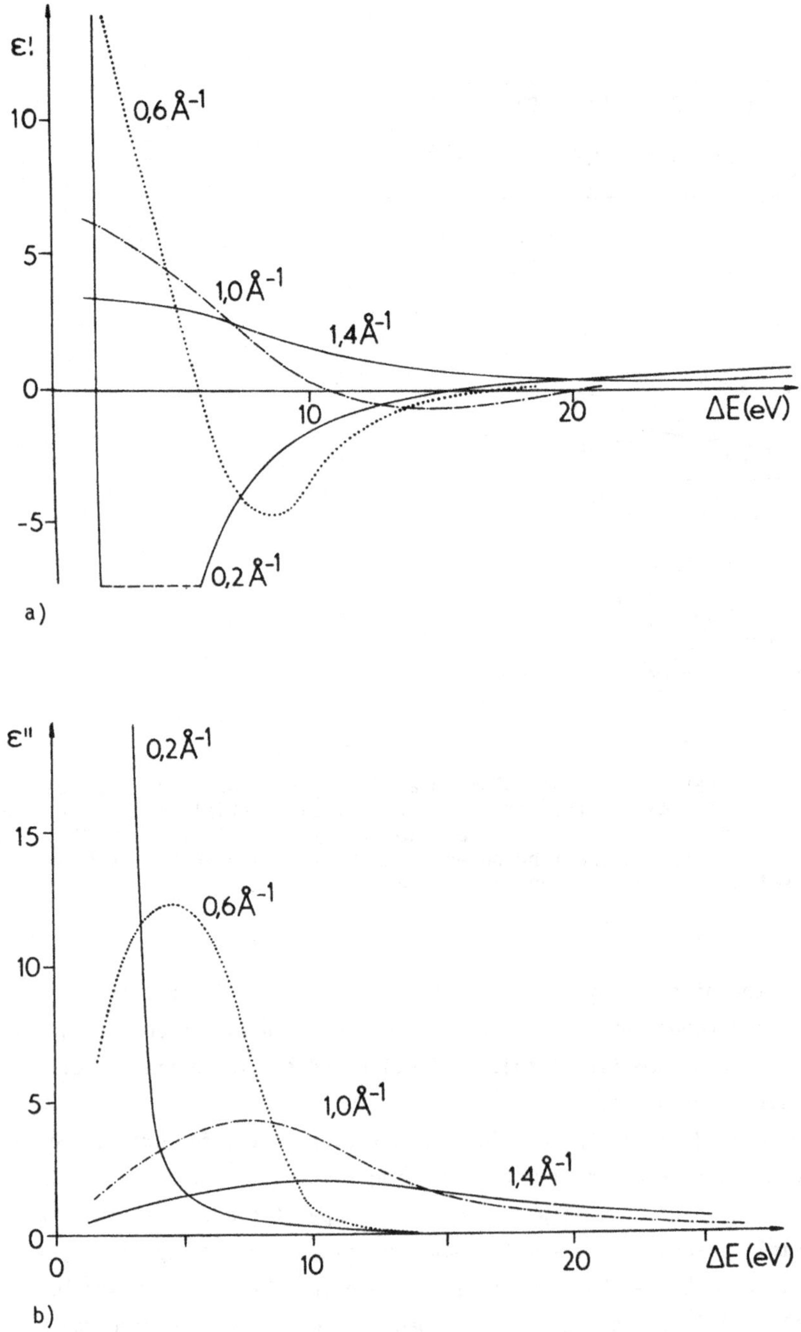

Fig. 1.3a,b. Lindhard's dielectric function with a damping constant of $\gamma = 1/\omega_p\tau = 0.033$ (case of Al)

1.5 Acoustic Plasmons

We have seen that the condition $\varepsilon(\omega,q) = 0$ applying the Lindhard dielectric function (1.16) in the approximation $q \ll q_F$ gives the dispersion relation of volume plasmons $\omega(q)$. The characteristic result is the finite plasma frequency ω_p for $q \to 0$.

The eigenfrequency remains finite although the wavelength λ goes to infinity with $q \to 0$; this can be explained by the fact that + and - charges lie on plane parallel sheets so that the restoring forces which determine the eigenfrequency are independent on the distance of the sheets or on λ (the number of charges assumed as constant). In case of surface plasmons the charges lie on rows, see Fig. 10.1, so that the restoring forces go to zero with $q \to 0$ and therefore the eigenfrequency too, see Fig. 10.3.

The question is discussed whether volume plasmons exist with an eigenfrequency which goes to zero with $q \to 0$ similar to the acoustic phonons. Such "acoustic" plasmons are expected under special conditions: a plasma shall consist of two components for example of light (mass m_1) and heavy electrons (mass m_2) with $m_1 \ll m_2$ and of equal density; thus the Fermi wave vectors $k_{Fi} = m_i v_{Fi}/\hbar$ (i: 1,2) are equal and $v_{F1} \gg v_{F2}$.

If we consider the LINDHARD function for small frequencies ω and q under the condition $v_{F1} \gg \omega/q \gg v_{F2}$ a dispersion relation of a mode of acoustic type follows

$$\omega = q\left(\frac{1}{3} v_{F1} v_{F2}\right)^{1/2} \tag{1.25}$$

with an imaginary term dependent on v_{F2}/v_{F1}. The above condition $v_{F1} \gg v_{F2}$ reduces the damping of this mode /1.1g/.

To illustrate this relation (1.25) we have to look at curve (2) of Fig. 1.2. Its slope is given by $\omega/q = v_F$ at $q \to 0$. In our case we have to draw two curves of tpye (2) with very different slopes v_{F1} and v_{F2}; between both these curves lies the curve ω/q of (1.25) representing the phase velocity of the acoustic mode, which fulfills the condition $v_{F1} \gg \omega/q \gg v_{F2}$. Lying in the continuum of single particle excitation for the light particles this mode is weakly damped corresponding to the value of v_{F2}/v_{F1}.

The two species of electrons do not oscillate in phase, so that the restoring forces compensate partially; they cancel with $q \to 0$ so that $\omega \to 0$.

Up to now experimental evidence has not yet been reported, since special conditions have to be fulfilled to produce them. The interest of such plasmons consists in the possibility of continuously changing the frequency which allows one to study their interaction with the bulk of the material in a similar way as the surface plasmons are used for surface studies.

2. The Dielectric Function and the Loss Function of Bound Electrons

Some important features of the collective behaviour of bound electrons, the dielectric function as well as the loss function, shall be considered in a simple approach.

2.1 Elementary Remarks. The Plasma Frequency

If an exterior field E acts on electrons bound with the eigenfrequency ω_n to the ions, we can describe their movements assuming that the local field acting on the electrons is equal to the exterior field

$$m\ddot{x} + \frac{m}{\tau}\dot{x} + m\omega_n^2 x = eE \tag{2.1}$$

Using the polarisation

$$P = (\varepsilon-1)\tilde{\varepsilon}_o E = eNx \tag{2.1}$$

with $\tilde{\varepsilon}_o$, (1.2), we obtain

$$\varepsilon(\omega) = 1 + \frac{Ne^2}{m\tilde{\varepsilon}_o}\,\frac{1}{\omega_n^2 - \omega^2 + i\omega/\tau} \tag{2.3}$$

N shall be the number of electrons per cm^3. In case of free electrons ($\omega_n = 0$) we get (1.4). If besides N_f free electrons, we have N_b electrons of eigenfrequency ω_n; the dielectric function becomes

$$\varepsilon(\omega) = 1 + \chi_f + \chi_b \tag{2.4}$$

with the contribution of free electrons

$$\chi_f = \frac{N_f e^2}{m\tilde{\varepsilon}_o}\,\frac{1}{-\omega^2 + i\omega/\tau} \tag{2.5}$$

and bound electrons

$$\chi_b = \frac{N_b e^2}{m\tilde{\varepsilon}_o} \frac{1}{\omega_n^2 - \omega^2 + i\omega/\tau} \tag{2.6}$$

The frequency of the volume plasma excitation with $1/\tau = 0$ comes out from $\varepsilon' = 0$ as

$$\tilde{\omega}_p^2 = \frac{\omega_p^2}{1 + \chi_b} \tag{2.7}$$

the changed plasma frequency.

Case (a): We assume that ω_n lies at higher frequencies

$$\omega < \omega_n \quad \text{and also} \quad \omega_p < \omega_n \tag{2.8}$$

so that we have with (2.6)

$$\tilde{\omega}_p^2 = \omega_p^2 - \frac{N_b e^2}{m\tilde{\varepsilon}_o} \frac{\omega_p^2}{\omega_n^2} \tag{2.9}$$

This means that the plasmon energy $\hbar\tilde{\omega}_p$ is reduced if the plasma frequency of the N_f electrons in the conduction band lies below the energy of the bound electrons. The tightly bound electrons move in antiphase to the free electrons reducing thus the eigenfrequency of the system, see Fig. 2.1 below. Example: the π plasmon of graphite is observed at 6.8 eV, but calculated at 12.6 eV.

Case (b): The reverse happens if the regarded frequencies ω are higher than the interband transitions. Then χ_f and χ_b have the same sign (the electrons, free and bound ones, are excited above their eigenfrequency) and with $\omega_n < \omega$ we obtain

$$\tilde{\omega}_p^2 = \omega_n^2 + \frac{N_b e^2}{m\tilde{\varepsilon}_o} \tag{2.10}$$

These results can also be derived from Fig. 2.1 as is seen without detailed discussion. It explains qualitatively the deviations of the measured value of ΔE_p from the calculated $\hbar\omega_p$, using the free electron concept.

We discuss as example an insulator ($N_f = 0$) with N_b bound electrons. The binding energy of these electrons $\hbar\omega_n$ shall be small compared to the plasma frequency $\hbar\omega_p$ (if these electrons would be free)

$$\omega_p^2 = \frac{N_b e^2}{m\tilde{\varepsilon}_o} > \omega_n^2 \tag{2.11}$$

Thus we write for the plasma frequency $\tilde{\omega}_p$

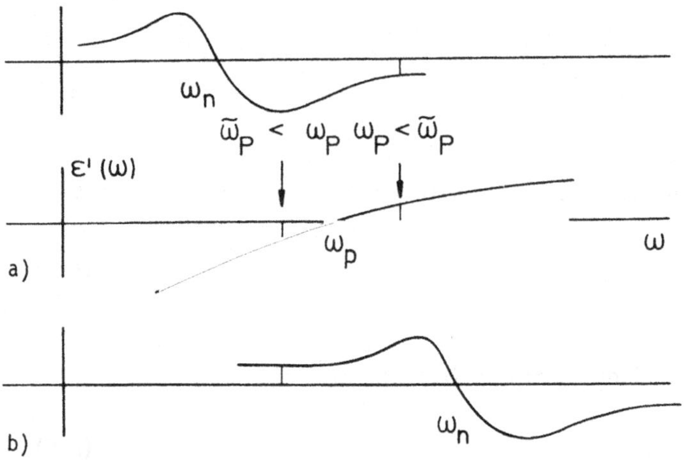

Fig. 2.1a,b. Free and bound electrons contribute to the dielectric function. (a) The eigenfrequency of the bound electrons ω_n lies below ω_p, the plasma frequency of the free electrons, so that the position $\varepsilon'(\omega) = 0$ is displaced to higher values: $\omega_p < \tilde{\omega}_p$. The reverse takes place for $\omega_n > \omega_p$, as is indicated in Fig. 2.1 (b)

$$\tilde{\omega}_p = (\omega_p{}^2 + \omega_n{}^2)^{1/2} \tag{2.12}$$

Here the plasma frequency of the loosely bound electrons is displaced to higher frequencies. This situation is realised in Si and Ge.

If we have a semiconductor or an insulator whose conduction band is doped with N electrons per cm^3, these electrons perform oscillations, volume - as well as surface oscillations, with a smaller frequency than that of metal electrons due to the low N. The influence of the lattice structure is described by (2.7). This situation can also be described by introducing into the relation (2.7) the effective mass $m^* = m_o\ (1 + \chi_b)$, so that

$$\tilde{\omega}_p{}^2 = \frac{Ne^2}{m^*\varepsilon_o} \tag{2.13}$$

2.2 The Loss Function

We assume that the crystal consists of electrons which are bound with the eigenfrequency ω_n corresponding to an interband transition $\hbar\omega_n$. It behaves then like an insulator with a gap of this value. Neglecting a q dependence, the dielectric function of (2.3) is valid and we can calculate the loss function $\mathrm{Im}(-1/\varepsilon)$ which shows a peak beyond the value $\varepsilon' \sim 0$, see Fig. 2.2. This peak means that a rather

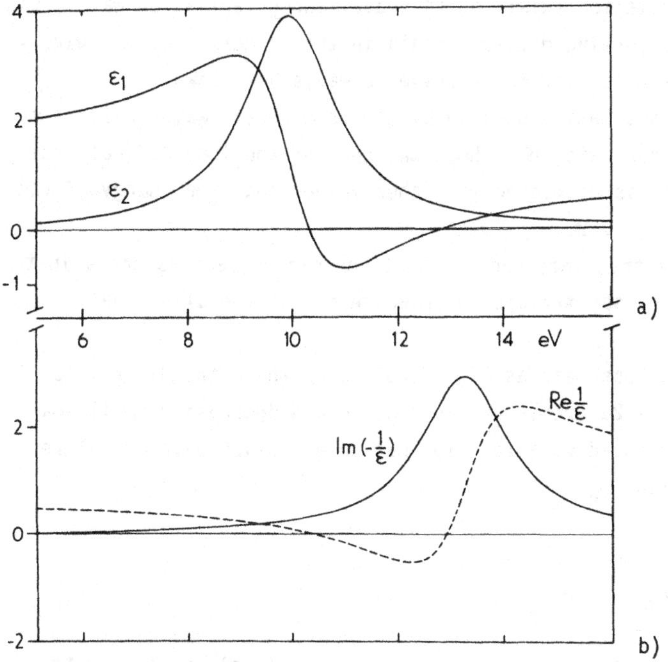

Fig. 2.2a,b. The loss function of a bound electron. (a) The dielectric function: ε_1 real part, ε_2 imaginary part. $\hbar\omega_n$ = 10 eV, $\hbar/\tau$ = 4 eV. (b) The function Im$(-1/\varepsilon)$ and Re$(1/\varepsilon)$ (dotted line) is calculated with $\varepsilon(\omega)$ of (a)

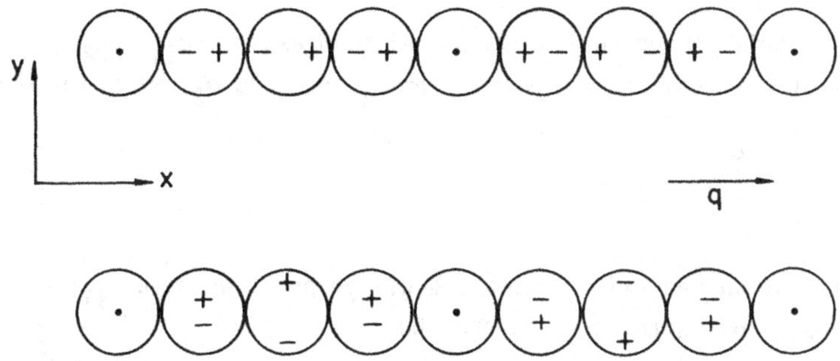

Fig. 2.3. Schematic presentation of the polarisation waves in crystals with bound electrons. Above: longitudinal wave (div P ≠ 0,P polarisation). Below: transverse wave (div P = 0)

pronounced longitudinal excitation exists as "a travelling polarisation wave", since the electrons are displaced forming dipole momenta in the direction of the wave vector. In Fig. 2.3 the longitudinal and the transverse waves are drawn.

Whereas the transverse oscillation excited by electromagnetic waves has its maximum amplitude at the peak of ε'' with $\hbar\omega = \hbar\omega_n$, the peak of the longitudinal excitation produced by electrons is displaced to higher values than the energy of the interband transition, see Fig. 2.2.

This result demonstrates that interband transitions can be seen as peaks in the loss spectrum. These types of spectra are observed, e.g., at the alkali halide crystals, see Fig. 5.1.

Vice versa: if we have a loss peak as Fig. 2.2b around which the dielectric function behaves as in Fig. 2.2a (real part increasing and decreasing small imaginary part) it can be attributed to a longitudinal polarisation wave correlated with the interband transition $\hbar\omega_n$.

2.3 Anisotropic Excitations

In case of an isotropic material the dielectric function $\varepsilon_\ell(\omega,q)$ as well as the loss function is independent of the direction of $\vec{q}$. However in an anisotropic crystal the dielectric function has the character of a tensor, so that the ε values are different in different crystal directions (also for $q \to 0$). The loss function becomes in this case /2.1/

$$\mathrm{Im}\,\frac{-1}{\sum_i\sum_j(\varepsilon_{ij}q_j)q_i} \tag{2.14}$$

If we transform the dielectric tensor to its orthogonal main axes: $\varepsilon(\varepsilon_{11}, \varepsilon_{22}, \varepsilon_{33})$ and call the components of $\vec{q}$ in this coordinate system q_1, q_2, and q_3, we can write

$$\mathrm{Im}\,\frac{-1}{\sum_i\varepsilon_{ii}q_i^2}\,, \tag{2.15}$$

see Fig. 2.4. In the case of an one-axial crystal, e.g., graphite we have $\varepsilon_{11} = \varepsilon_{22} = \varepsilon_\perp$ and $\varepsilon_{33} = \varepsilon_\parallel$ (parallel to the c axis) and thus

$$\varepsilon(\omega,\Theta) = \varepsilon_\perp\sin^2\Theta + \varepsilon_\parallel\cos^2\Theta \tag{2.15a}$$

with Θ the angle between the c axis and $\vec{q}$. Thus one can follow how the loss spectrum develops from $\vec{q}\parallel c$ to $\vec{q}\perp c$ axis /1.1d/.

By turning the crystal in respect to the incoming electron beam it is possible to bring the direction of the vector $\vec{q}$ parallel to the axis ε_{ii} of the dielectric tensor, so that the loss function

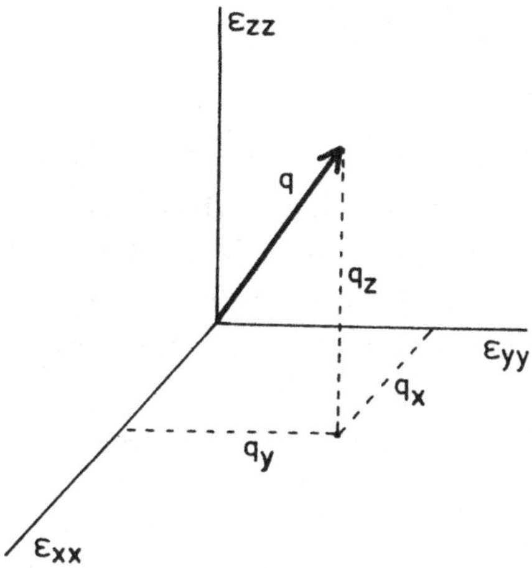

Fig. 2.4. The wave vector q and the main axes of the dielectric tensor ε_{ii}

$$\mathrm{Im}\,\frac{-1}{\varepsilon_{ii}q_i^2} \tag{2.16}$$

results. Thus the component ε_{ii} of the tensor can be measured. As an example the loss function for $q \parallel c$ axis and $q \perp c$ axis in graphite are reproduced in Fig. 5.6.

In (2.14-16) we have included in the loss function the wave vector q; in general we call $\mathrm{Im}(-1/\varepsilon)$ without $1/q^2$ the loss function.

It shall be added that in an anisotropic crystal volume modes with longitudinal (transverse) electric fields exist if the vector q lies parallel (perpendicular) to one of the main axes of the dielectric tensor $\hat{\varepsilon}$. In intermediate directions the volume modes are a mixture of these modes with fields of transverse and longitudinal components. In isotropic crystals longitudinal and transverse volume modes are separated in all directions /2.1a/.

2.4 The Dielectric Function $\varepsilon_{GG'}(\omega,q)$

The considerations of Sect. 2.2 are useful for a simple discussion of the experimental results. The question arises whether we can derive the dielectric function from more fundamental considerations.

Calculation of $\varepsilon(\omega,0)$ *with* $q = 0$. If one assumes that the exterior field (electrons or light) transfers its energy to the solid by exciting interband transitions in the one-electron approximation it is possible on the basis of the quantum theory of interband transitions to calculate the number of transitions per unit time and per unit volume due to the exterior field. Thus one obtains, see /2.2/

$$\varepsilon''(\omega,0) = \frac{4\pi^2e^2}{m^2\omega^2}\sum_{n,n'}\int_{BrZ}\frac{2}{(2\pi)^3}\,d^3k\,|\vec{e}\cdot\vec{M}_{nn'}(\vec{k})|^2\cdot\delta\left(E_n(\vec{k}) - E_{n'}(\vec{k}) - \hbar\omega\right) \tag{2.17}$$

with $M_{nn'}(k)$ the matrix element for the transition from the filled level n of one band with a given k value to an empty band n' with the same k value, within the Brillouin zone. This means that no momentum is transferred to the excited electron ($q = 0$). This important relation connects optical properties with the band structure of the solid. With the help of the Kramers-Kronig relation $\varepsilon'(\omega,0)$ can be calculated too.

Under the assumption that $M_{nn'}$ can be regarded as independent on k the above equation can be written

$$\varepsilon''(\omega,0) = \frac{\text{const}}{\omega^2}\cdot|M_{nn'}|^2\cdot J_{nn'}(\hbar\omega) \tag{2.18}$$

with $J_{nn'}$ the joint density of the occupied and the empty states which have an energy difference of $\hbar\omega$. Successful calculations of $\varepsilon(\omega)$ are reported on Ge /2.3/, Graphite /2.4/, Al /2.5/, and other substances.

Calculation of $\varepsilon(\omega,q)$ with finite q. We shall see that observations with finite q or observations of energy losses at finite scattering angles θ give important information of the properties of the plasmons. A dielectric function for finite q has been obtained first by Lindhard for free electrons as we discussed in Chap. 1. An important step beyond has been done by including the effect of the periodic potential of the lattice: In a series of publications this microscopic dielectric function $\varepsilon_{GG'}(\omega,q)$ has been elaborated by EHRENREICH and COHEN /2.6/, ADLER /2.7/, and WISER /2.8/ who introduced the band structure effects and the local field corrections. The dielectric function looks then as follows

$$\varepsilon_{GG'}(\omega,q) = \delta_{GG'} - \lim_{\tau\to\infty}\frac{4\pi e^2}{\Omega\cdot|\vec{q}+\vec{G}|\cdot|\vec{q}+\vec{G}'|}\sum_{nn'k} L_{nn'k} \tag{2.19}$$

$$L_{nn'k} = \frac{f_0(E_{n'}(k+q)) - f_0(E_n(k))}{E_{n'}(k+q) - E_n(k) - \hbar(\omega+i/\tau)}\,\eta_G^*\,\eta_{G'} \tag{2.20}$$

$$\eta_G = \frac{1}{V_c}\int_{cell} u_{k,n}^*(r)\,e^{-iGr}\,u_{k+q,n'}(r)\,d^3r \tag{2.21}$$

G, G': reciprocal lattice vector; Ω: volume of the crystal = NV_c; V_c: volume of the cell; n, n': band indices; k: wave vector of the electrons; f_0: Fermi distribution; $u_{k,n}(r)$: Bloch function.

The energy levels $E_n(k)$ are those obtained from the band structure and not from free electron gas theory.

This function represents a matrix which elements are determined by the vectors G, G'. It goes over into a function derived by EHRENREICH and COHEN /2.6/, if we put G = G' = 0, but retain the band structure (Bloch function, interband levels)

$$\varepsilon_{00}(\omega,q) = 1 - \lim_{\tau\to\infty} \frac{4\pi e^2}{\Omega q^2} \sum_{nn'k} \frac{f_o(E_{n'}(k+q)) - f_o(E_n(k))}{E_{n'}(k+q) - E_n(k) - \hbar(\omega+i/\tau)} \cdot |M_{k,n;k+q,n'}|^2 \tag{2.22}$$

$$M = \frac{1}{V_c} \int_{cell} u^*_{k,n}(r) \cdot u_{k+q,n'}(r)d^3r$$

The influence of the periodic lattice potential on the field inside the elementary cell or the local field correction is thus dropped. Replacing further the Bloch function $u_{nk}(r)$ by the value 1 and $E_n(k)$ by $\hbar^2k^2/2m$ we get the Lindhard expression. For $q \to 0$ the dielectric function discussed in (2.17) is obtained.

The dielectric function $\varepsilon_{GG'}$, containing the G vectors produces an electric polarisation field which varies rapidly in space, even if the perturbing exterior field D varies only slowly. This "local field correction" is important and improves the agreement between calculated and observed values considerably, see Fig. 6.5.

This generalisation of the dielectric function has some consequences on how the plasma frequency and the loss function are to be determined. The condition $\varepsilon(\omega,q) = 0$ used earlier in Chap. 1 yielding the dispersion relation has to be replaced by

$$\text{Det}\ \varepsilon_{GG'}(\omega,q) = 0 \tag{2.23}$$

where Det means the determinant of the $\varepsilon_{GG'}$ elements. This condition results from the Fourier transform

$$D_G(\omega,q) = \sum_{G'} \varepsilon_{GG'}(\omega,q) \cdot E_{G'}(\omega,q) \tag{2.24}$$

In order to get a finite $E_{G'}$ at $D_G = 0$ (no exterior perturbation) the determinant of the system of linear equations (D) has to be zero.

Introducing the dielectric function into the loss function the value of ε^{-1} has to be determined. The result is

$$\varepsilon^{-1}(\omega,q) = \left[\varepsilon_{GG'}(\omega,q)\right]^{-1}_{00}\ , \tag{2.25}$$

i.e., the first element G = G' = 0 of the inverted matrix $\varepsilon_{GG'}$ has to be calculated;

its imaginary part represents the loss function Im $\varepsilon^{-1}(\omega,q)$. Inverting the matrix $\varepsilon_{GG'}$ has the consequence that ε^{-1} involves all the G processes. In addition to the structure influence local field effect comes thus into play when one calculates the line width, its q dependence, the dispersion of the line position, and their anisotropy. These quantities are just those which can be measured with fast electron loss spectroscopy rather exactly; the results thus represent a good test for the theoretical considerations, as we shall see in Chap. 6 and Sect. 7.4. It should be added that this treatment is an approximation too, since electron correlation is neglected. Taking these effects into account may lead to only small corrections but it is very difficult to estimate their importance accurately.

If the off-diagonal elements of the matrix $\varepsilon_{GG'}$ can be neglected, then Det $\varepsilon_{GG'} = 0$, (2.23), reduces to the product of $\varepsilon_{GG'} = 0$. Each factor of this product set equal to zero gives a piece of the dispersion curve; put together they represent the total dispersion curve.

It has been suggested that at the zone boundaries q = G splitting of the dispersion curve should be expected /2.9-11/, see Fig. 2.5. If we assume that only one band G exists, the determinant $\varepsilon_{GG'}$ reduces to

$$\begin{vmatrix} \varepsilon_{00}(\omega,q) & \varepsilon_{0G}(\omega,q) \\ \varepsilon_{G0}(\omega,q) & \varepsilon_{GG}(\omega,q) \end{vmatrix} = 0 \tag{2.26}$$

which leads to a quadratic condition for $\omega_p(q)$. If the value of q_c, the cutoff vector, lies above G/2 there should be a chance to observe this splitting; in general however the width of the losses will have increased so much due to the q dependence of $\Delta E_{1/2}$ (see Chap. 6) that it will be difficult to find this effect.

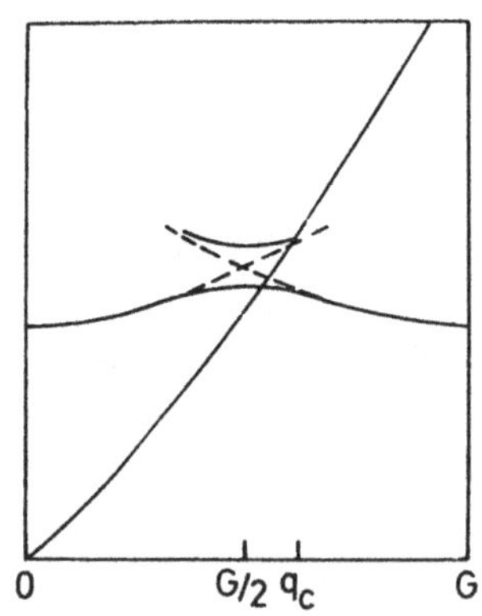

Fig. 2.5. Scheme which illustrates the formation of two plasmon bands (solid lines) due to Bragg reflection at q = G/2. The curve starting at q = 0 and ω = 0 is identical with curve (2) of Fig. 1.2

3. Excitation of Volume Plasmons

After having discussed the different properties of these longitudinal eigenmodes of the electron gas we are interested to know how they can be excited. We shall see that we have different means: electrons fast as well as slow, X-rays, and under special conditions photons which can excite this mode at a boundary of a plasma.

3.1 Excitation by Electrons

Probability of excitation. The first observations of such energy losses in transmission of thin films were made with 7.5 keV electrons which suffer inelastic collisions with the crystal electrons, RUTHEMANN /3.1/. These losses ΔE of the inelastically scattered electrons inform on the properties of these excitations. Their position, halfwidth, as well as intensity are strongly correlated to the dielectric behaviour or the polarisability of the solid described by the complex dielectric function $\varepsilon(\omega,q)$. This can be understood as follows: The primary electrons (charge e) entering the electron gas with the velocity v represent a current density $j = e \cdot v \cdot \delta(x-vt)$; they polarise the medium, so that the electric field E_{pol} resulting from this polarisation interacts with the incoming electron. It acts against the electron, so that it loses the energy dA per path length ds

$$-dA = |e| \;\vec{E}_{pol} \;\frac{\vec{v}}{|v|}\; ds. \tag{3.1}$$

E_{pol} can be calculated from Maxwell's equations. If we define the elementary probability

$$\frac{\partial^3 W}{\partial \Delta E \partial^2 \Omega}$$

for an electron losing ΔE per path length ds and being scattered into the solid angle $d^2\Omega$ by a deflection θ we have

$$-dA = \iint \Delta E \cdot \frac{\partial^3 W}{\partial \Delta E \, \partial^2 \Omega} \, d\Delta E \, d^2\Omega. \tag{3.2}$$

With the knowledge of E_{pol}, which needs a more extended calculation see /1.1b/ and by comparison of (3.1) with (3.2) we obtain

$$\frac{\partial^3 W}{\partial \Delta E \partial^2 \Omega} = \frac{1}{(e\pi a_0)^2} \frac{D}{q^2} \cdot \mathrm{Im}(-1/\varepsilon) = \frac{1}{(e\pi a_0)^2} \cdot \frac{D}{q^2} \cdot \frac{\varepsilon''}{|\varepsilon'|^2+|\varepsilon''|^2}, \tag{3.3}$$

the electron having passed a length D. Here the thickness has to be so small that no multiple processes take place. For multiple processes see Chap. 4. If we take into account the q dependence of ε we obtain the same relation with $\varepsilon(\omega,q)$. The function $\mathrm{Im}(-1/\varepsilon)$ is called the loss function, already mentioned in Sect. 2.2.

The quantum mechanical calculation /3.2/ has the same result for the excitation probability. It gives in addition some interesting relations: If $n(\vec{r},t)$ is the electron density at $\vec{r}$ and t, the (normalized) electron density-density correlation function is written

$$G(\vec{r},t) = \frac{1}{n^2} < n(\vec{r},t) \cdot n(0,0) > . \tag{3.4}$$

Its Fourier transform results as

$$S(\omega,q) = \frac{n}{2\pi} \int_{-\infty}^{+\infty} d^3\vec{r} \int_{-\infty}^{+\infty} dt \; e^{-i(\vec{q}\cdot\vec{r}-\omega t)} \left| G(\vec{r},t)-1 \right| . \tag{3.5}$$

Now this dynamical structure factor $S(\omega,q)$ is related to the differential excitation probability by

$$\frac{\partial^3 W}{\partial \Delta E \partial^2 \Omega} = \mathrm{const} \cdot \frac{1}{q^4} \cdot S(\omega,q). \tag{3.5a}$$

It is however not necessary to know this S function, since it is related to the loss function as follows /3.2/

$$S(\omega,q) = \frac{q^2}{4\pi^2 e^2} \cdot \mathrm{Im}\{-1/\varepsilon(\omega,q)\}. \tag{3.5b}$$

This result which makes the calculation of the probability much easier, is identical with (3.3), the classical result.

These equations demonstrate that the inelastic scattering of fast electrons passing a free electron gas is produced by the correlation of the electron densities; these are strong in the plasma oscillations of frequency ω_p.

The calculation of the function G(r,t) is an actual problem of the Fermi liquid theory.

The loss function reduces to ε'' if the medium is dense as in a gas ($\varepsilon' \sim 1$ and $\varepsilon'' \ll 1$).

As we see the loss probability is dependent on two factors, the wave vector q and the term $\mathrm{Im}(-1/\varepsilon)$. The loss function will be treated later at different occasions. Here the q dependence shall be discussed. The transferred momentum $\hbar q$ can be correlated in the case of fast electrons or $\Delta E \ll E_0 = \hbar^2/2m \cdot k_{el}^2$ with the scattering angle of the fast electrons θ using the momentum conservation, see Fig. 3.1

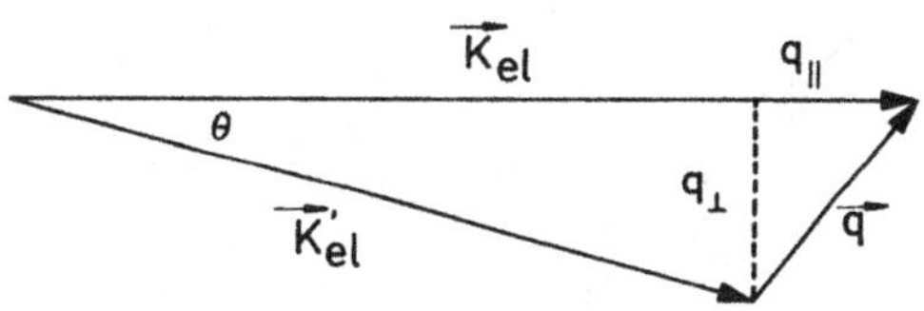

Fig. 3.1. Conservation of the momentum of the incoming electrons ($\hbar\vec{k}_{el}$), of the scattered electrons ($\hbar\vec{k}'_{el}$) and of the excited plasmon ($\hbar\vec{q}$) (or of an excitation in general)

$$\hbar\vec{q} = \hbar\vec{k}_{el} - \hbar\vec{k}'_{el} \tag{3.6}$$

and energy conservation

$$\Delta E = \hbar\omega = \frac{\hbar^2}{2m}\left(k_{el}^2 - k'^2_{el}\right). \tag{3.7}$$

We obtain

$$q^2 = k_{el}^2\,\theta^2 + \left(\frac{\omega}{v}\right)^2 = q_{\perp}^2 + q_{\parallel}^2 \tag{3.8}$$

for $v/c \ll 1$. $q_{\parallel}$ can be written

$$q_{\parallel} = \theta_{\Delta E} \cdot k_{el} \tag{3.9}$$

with

$$\theta_{\Delta E} = \Delta E/2E_0. \tag{3.10}$$

Since $\theta_{\Delta E}$ has in general a small value, e.g., with $\Delta E = 10$ eV and $E_0 = 50$ keV it is around 10^{-4} rad, the transferred momentum can be written for $\theta > \theta_{\Delta E}$

$$q = q_{\perp} = k_{el} \cdot \theta, \tag{3.11}$$

see Fig. 3.1. We see that the probability (3.3) remains constant for $\theta < \theta_{\Delta E}$ and decreases for $\theta > \theta_{\Delta E}$ as $1/\theta^2$, see Fig. 3.2. This strong decrease of the intensity at large q values leads to difficulties to observe intensities at high q values.

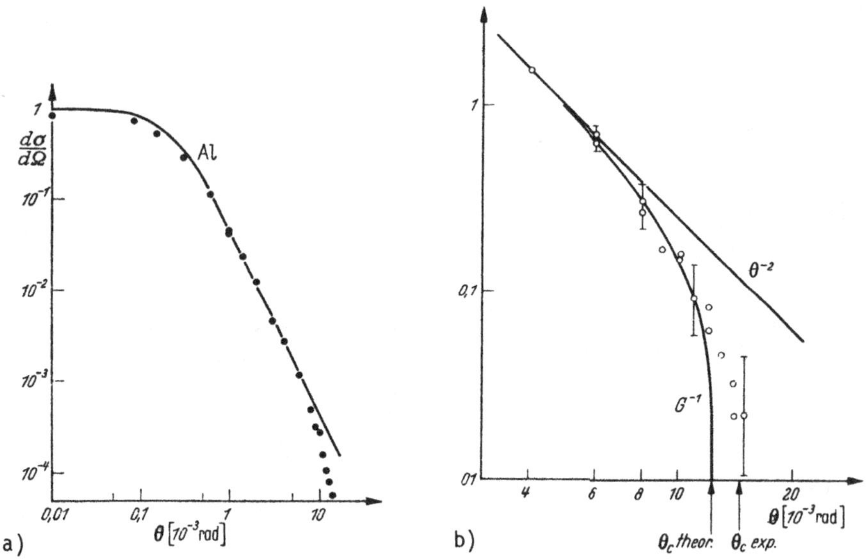

Fig. 3.2. (a) The measured angular dependency of the (relative) intensity of the 15 eV loss in Al compared with the calculated curve in a logarithmic scale. The theoretical curve is obtained by folding the angular distribution of the primary beam, measured at the no loss beam, with the excitation probability (3.3). Both curves, the experimental and the theoretical, fitted at the angle = 2 mrad, are in good agreement /3.3/.
(b) The same dependence as Fig. 3.2a, at angles $\theta \sim \theta_c$. The function $G^{-1}(q,q_c)$ describes the measurements rather well; q_c (exp) = 1.5 Å^{-1}, q_c (theor) = 1.3 Å^{-1}

The intensity decreases somewhat quicker than θ^{-2} at values of $\theta \sim \theta_c$, see Fig. 3.2; this effect can be explained by the q dependence of the dielectric function, see Fig. 7.14 and /3.4/. Another way to describe the quicker decrease of the inelastic intensity is to introduce into (3.3) a factor $G^{-1}(q,q_c)$ /3.5/ which fits the observed intensity drop too.

The cutoff wave vector q_c itself comes out in good agreement with the calculated values having in mind that observed and theoretical q_c values can be determined only approximately. For example, the experimental values of q_c for Al, Mg, Sb, are 1.5 $Å^{-1}$ /3.3/, 1.2 $Å^{-1}$ /3.3/, 1.0 $Å^{-1}$ /3.6/, and the calculated ones are 1.3, 1.1, and 1.2 $Å^{-1}$.

The direction of the scattered electron deviates in general only by small angles θ from the primary direction; the vector $\vec{q}$ however makes an angle η with it, which can be rather large. The angle η is derived from Fig. 3.1 and 3.3

$$\mathrm{tg}\eta = q_{\perp}/q_{\parallel} = \theta/\theta_{\Delta E}. \tag{3.12}$$

At $E_0 = 50$ keV and $\Delta E = 10$ eV already a small deviation of $\theta \sim 10^{-3}$ rad gives $\eta > 80^{\circ}$, so that the plasmon runs nearly perpendicular to the primary electron beam into the crystal.

Equation (3.12) shows that η is dependent on ΔE. In anisotropic crystals with $\varepsilon(\omega,\vec{q})$ one has to take care of this change of the direction of $\vec{q}$ with ΔE /3.7/. As we see the knowledge of the spectral dependence of the dielectric function is needed as well as its spatial dispersion (q dependence) to calculate the electron energy loss function.

For small vectors q (small compared with the extension of the Brillouin zone) $\varepsilon(\omega,q)$ can be replaced by $\varepsilon(\omega,0)$ the dielectric function obtained from measurements with transverse electric fields or light. This is an important result of the theory verified by many experiments that the electron energy loss spectrum for $q \sim 0$ can be deduced from the knowledge of the spectral dependence of the optical constants /1.1d/. We can say that the excitation by electrons is nearly identical with that by light, if the momentum transferred to the crystal electrons is small, or if we have to do with nearly vertical transitions.

If the electrons pass an anisotropic (uni- or biaxial) crystal the excitation probability becomes a function of the direction of $\vec{q}$ as has been shown for the loss function, see (2.14). The intensity relations are elaborated in /3.7a/.

Double processes. The above introduced probability for plasmon excitation describes the production of a single plasmon; however there exists the possibility to create multiple plasmons as a consequence of a nonlinear response, e.g., a double plasmon loss $2\hbar\omega_p$ in one collision. In contrast to the multiple processes described in Sect. 4.2 which consist of several single processes taking place one after the other these multiple plasmons are created in one interaction process. It is characteristic that this excitation probability is thickness independent, which gives theoretically a method to distinguish the two processes. To have an idea of the importance of this process, we consider the total number of electron losses in a loss spectrum at the energy position $2\hbar\omega_p$; the fraction f of the energy losses being produced in one collision has been calculated as /3.8/

$$f = \left\{1 + \frac{1}{0.02\, r_s^2} \cdot \frac{D}{\Lambda}\right\}^{-1} \tag{3.13}$$

with D the thickness of the film, Λ the mean free path length [see (4.17)], and $r_s = (4\pi/3)^{1/3} \cdot n^{1/3} \cdot a_0)^{-1}$ the mean distance of two electrons in units of the Bohr radius. With $r_s = 2$ (Al) we obtain with $D = \Lambda$ from (3.13): $f = 0.08$ or 8 % of the double losses can be ascribed to one collision process. With D = 0 all losses are produced by this way. We can check this result by counting the number of losses at $2\hbar\omega_p$. The contribution of double losses due to multiple processes can be calculated (Sect. 4.1) so that we obtain from the difference the number of events due to double plasmons produced in one collision. The experiments performed at thin Al and Si films however did not allow one to remove the double plasmon loss produced in two steps accurately enough to demonstrate the existence of the double plasmon created in one process /3.9/. In another type of experiment it is believed that the theoretical prediction has been verified /3.10/; further work is needed to establish this result.

Considerations have been reported on the possibility of forming a bound state of two plasmon excitations in an electron gas. The investigation of the interaction of two plasmons via the exchange of an electron-hole pair results in an attraction so that it should be possible to excite biplasmons in special solids /3.10a/.

The electrons passing the condensed material excite besides the volume excitations in general surface modes or surface polaritons, guided light modes, and light (Cerenkov radiation). These modes and their excitation probability are treated in Chaps. 9 and 10.

Transmission and reflection method. An electron loss spectrograph to observe the loss spectrum by transmission of thin solid films with fast electrons ($E_0 \geq 30$ keV) is drawn schematically in Fig. 3.3. The energy of the electrons having lost ΔE and being scattered in this inelastic collision by an angle θ which is of the order of some 0.1 up to some 10 mrad is analysed in the spectrograph A; the intensity of the electrons having lost ΔE is measured in the detector D in general by counting the number of electrons. In this way the dispersion $\Delta E(\theta)$ can be studied, the dependence of the loss probability of θ, etc.. For more details of the experimental arrangement see Chap. 11. Examples of loss spectra are shown in the following chapter 5. Most of the experiments with fast electrons are performed by transmission of thin films; their thickness is in general comparable with the mean free path length for energy losses. This is about 10^3 Å for 50 keV electrons. Electrons of higher energies, 300 keV or more, can penetrate much thicker films and allow the investigation of specimens which can be prepared more easily. The mean free path length of 1000 keV electrons is about 4 times longer in Al than that of 50 keV electrons, see Sect. 4.2.

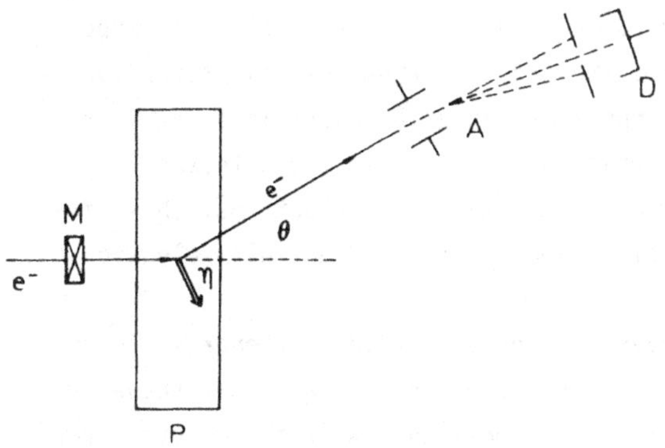

Fig. 3.3. Scheme of an electron energy loss spectrometer to observe excitations of momentum $\hbar q$ by electrons scattered in the solid plasma P. The energy loss ΔE of the electrons deflected by an angle θ is measured with an energy analyser (A) and the detector (D). M is a monochromator

Besides the transmission method there exists the reflection method; here the electrons are reflected at the boundary of a thick crystal. The intensity of the volume losses depends on the penetration depth of the electrons into the bulk, whereas the probability of surface plasmon excitation is a sensitive function of the angle of incidence, see Chap. 10. For slow electrons, $\sim$ 100 eV and less, such as those used in LEED experiments, the reflection method is the only one applicable for ELS of solids, since the mean free path length of inelastic scattering is of the order of about 10 Å due to the strong interaction of the electrons with the solid. ELS experiments with slow electrons in reflection have been made rather early /3.11/; Fig. 3.4 shows the scheme of the old device and explains the mode

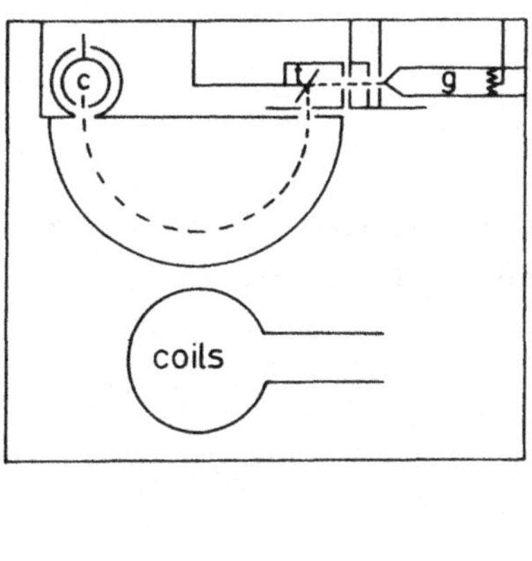

Fig. 3.4. Scheme of the apparatus of loss spectroscopy of slow electrons used by RUDBERG. The electrons emitted from the electron gun (g) hit the target (t) and are scattered into the entrance slit of the magnetic analyser. The magnetic field perpendicular to the plane of drawing is indicated by the coils. After the electrons had passed the analyser they are detected by the Faraday cage C /3.11/

of operation. In recent years the method has been developed especially in respect to high energy resolution, see Fig. 11.10. The great interest of this technique is due to its high sensitivity to surface properties since the first layer of a crystal contributes to a large extent to the intensity scattered into the detector.

Reflection experiments have also been performed with fast electrons. These investigations had the aim to study the surface and volume plasmons with electrons at a grazing incidence, see Chap. 10.

In recent years polarised electrons have been used to study low-energy electron diffraction; spin polarisation effects are observed which indicate that these are sensitive to structural changes /3.12/. Similar experiments will be important for the inelastic interaction of electrons too.

Polarisation wake. If a charged particle of velocity v passes a material with a plasma of nearly free electrons (plasma frequency ω_p) a spatially oscillating, linear charge density distribution is produced behind the particle (N. Bohr, 1948). Its wavelength has the value $2\pi v/\omega_p$ which amounts to about 20 Å in the case of heavy particles (e.g. three proton masses, see below) of 1 MeV and $\hbar\omega_p$ = 15 eV (Al). The periodic potential associated with this electronic polarisation "wake" can interact with charged projectiles running not too far from it. The electric field, the derivative of this potential at the particle position, makes the well known stopping power.

Experimentally the influence of this potential variation has been observed with heavy ions: $(HHe)^+$ ions which cross a thin foil and lose their electrons decompose by Coulomb explosion into H^+ and He^+. The distribution of the protons measured in energy and angle showed a difference depending on the orientation of the molecule $(HHe)^+$: If the proton leaves the molecule so that it runs in the track of the $(HHe)^+$-particle, emission in the backward direction, it is influenced by the wake potential and is deflected towards the beam direction. Emission in the foreward direction has a smaller effect. Thus the distribution of the counted protons shows an asymmetric behaviour due to this wake potential /3.12a/.

3.2 Excitation with X-Rays

Another method uses X-rays to excite the crystal electrons. It is known that X-rays are inelastically scattered by single electrons and transfer the momentum

$$\hbar q = (2\hbar\omega/c)\sin\theta/2 \qquad (3.14)$$

and the energy

$$\Delta E = \frac{\hbar^2}{2m} q^2 = 2\frac{(\hbar\omega)^2}{mc^2} \cdot \sin^2\theta/2 \qquad (3.15)$$

with θ the scattering angle of the X-rays. In the case of an electron plasma as scatterer, it is possible, if the dispersion relation for volume plasmons $\omega(q)$ with $q < q_c$ is fulfilled, that ΔE and $\hbar q$ are transferred to a collective of electrons, see Fig. 3.5. First results had been reported in /3.13,14/. An example of such an X-ray loss spectrum is shown in Fig. 3.6 /3.15/.

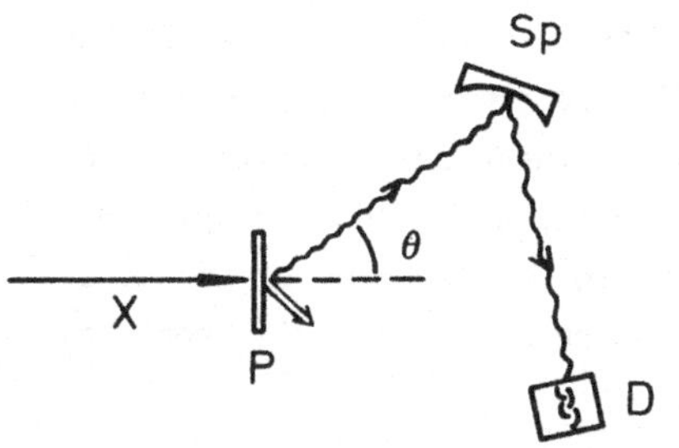

Fig. 3.5. Scheme of the arrangement to observe the excitation of plasmons by X-rays scattered by an angle θ inside the solid plasma (P). The energy loss of the X-rays is measured by the crystal spectrometer (Sp) and the detector (D)

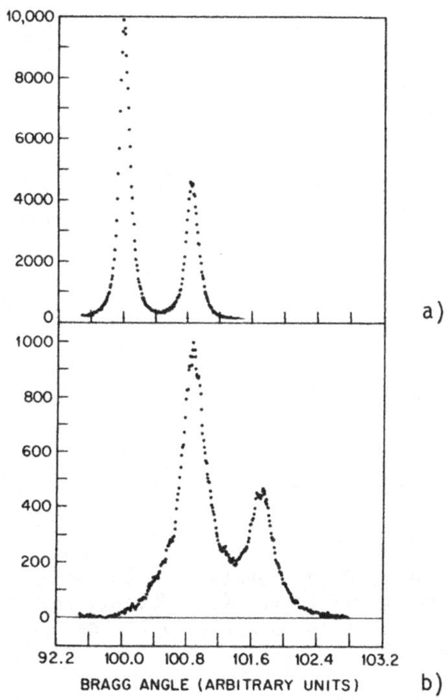

Fig. 3.6a,b. X-ray loss spectra obtained by sweeping double-crystal spectrometer. (a) Elastically scattered X-rays (Cu $K_{\alpha1}$, $K_{\alpha2}$). (b). Be plasmon loss lines from $K_{\alpha1}$ and $K_{\alpha2}$ at $q = 1\ \text{Å}^{-1}$, $q \parallel c$ axis after removal of elastic components /3.15/

This inelastic line had not been detected in earlier experiments since the Compton line has been studied in general at large angle θ ($\sim 90^0$) and with hard X-rays in order to obtain important changes of $\Delta\lambda/\lambda$, the energy resolution of the apparatus being limited. The plasmon experiments have to be performed at low q values ($\lesssim q_c$) or small θ values to measure energy changes ΔE of about 10 eV, see (3.15)

The probability of such an energy loss has been calculated already rather early /3.2/ for $q < q_c$

$$\frac{\partial^3 W}{\partial\hbar\omega\partial^2\Omega} = \text{const} \cdot \frac{n \cdot D}{4\pi e^2} \cdot q^2 \cdot \text{Im}\{-1/\varepsilon(\omega,\vec{q})\} \tag{3.16}$$

with const (the Thomson factor): $(e^2/mc^2)^2\ (\vec{e}_1 \cdot \vec{e}_2)^2$, $\vec{e}_{1,2}$ the polarisation vector before and after the scattering. Here as in (3.3) the loss function given by the imaginary part of the reciprocal dielectric function determines the dependence of the excitation probability on ω and $\vec{q}$. In contrast to the case of electrons the excitation probability increases with q up to q_c. This dependence is displayed in Fig. 3.7. Here the inelastic peak height of Be is drawn as function of the scattering angle, which increases up to θ_c ($\theta_c \cong 16^0$) whereas the inelastic peak height excited by electrons decreases as Fig. 3.2a shows.

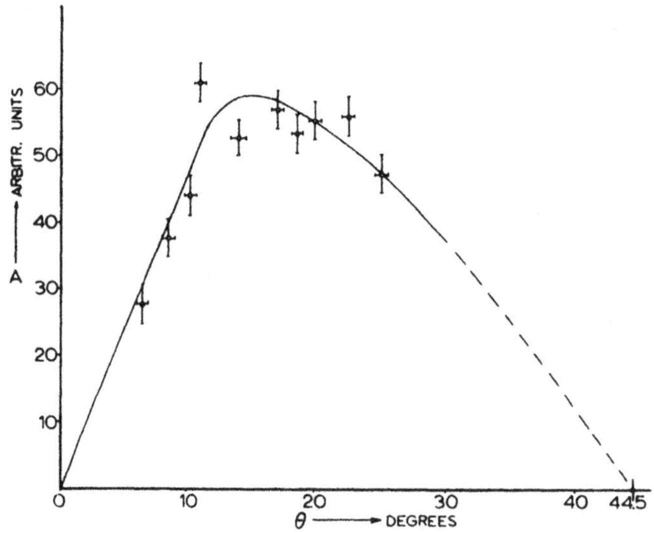

Fig. 3.7. The observed height of the plasmon peak of Be as function of the scattering angle; it increases with θ up to about the cutoff vector (q_c) /3.16/. For larger q the peak height goes down since the loss function, see Fig. 7.14, decreases more rapidly than q^2 increases. ΔE = 18.5 eV

3.3 Excitation by Light

We have seen that hard photons can excite plasmons in a volume. What about light? The wave vector of visible light is nearly 10^3 times smaller than that of X-rays. Since the probability of exciting plasma oscillations by photons is proportional to q^2, the probability is too small for visible light. However, there exists a possibility to excite volume plasmons with light via the surface of the crystal: the electric field of light falling on the boundary of a solid under an angle θ against the normal has a component E_n normal to the surface (direction z)

$$E_z^{(1)} = E_n^{(1)} = E_o \cos\theta . \tag{3.17}$$

It is regarded usually as discontinuous due to a surface charge layer of a density

$$\sigma = E_n^{(2)} - E_n^{(1)} ; \tag{3.18}$$

the indices 1 and 2 refer to the exterior and the interior of the material. This surface charge is regarded as a δ function-like distribution in the surface. A more realistic picture however would be a space charge of finite density and its continuous transition in the boundary from $E_n^{(2)}$ to $E_n^{(1)}$. Due to this changed distribution a finite gradient of the electron density n exists and leads to an additional force ($\sim$ dn/dz), which pushes the metal electrons along the z direction according to the frequency of the light and thus produces longitudinal density oscillations of the electrons or volume plasmons. The frequency of the light ω has to be $\geq \omega_p$ if real wave vectors q shall exist as follows from the dispersion relation, see (1.18):

$$\sqrt{\frac{3}{5}}\, v_F \cdot q = \sqrt{\omega^2 - \omega_p^2} . \tag{3.19}$$

The excitation of the plasmons by light should be observable by a decrease of the reflectivity R of the light at ω_p since photon energy is fed into plasma oscillations. The effect however is very small and not measurable; it can be strongly enhanced if one replaces the semi-infinite metal by a thin metal film, so that the density wave is reflected and standing waves are produced /3.16a,17,18/. When the frequency of the light is changed at a given film thickness resonance of R occurs if the dispersion relation is fulfilled. The wave vector q belonging to this resonance frequency is given by $q = n\pi/D$, n = 1,3,5 and can be derived from the film thickness D which had to be well known. The experiments are difficult to perform since smooth and very thin films (because of the rather strong damping of the plasmons) have to be prepared. Therefore this method has been applied only in special cases as at Ag films /3.19/ and at K films /3.20/. Figure 3.8 demonstrates these resonances at Ag films of different thicknesses. For the dispersion relation derived from these measurements see Fig. 7.5.

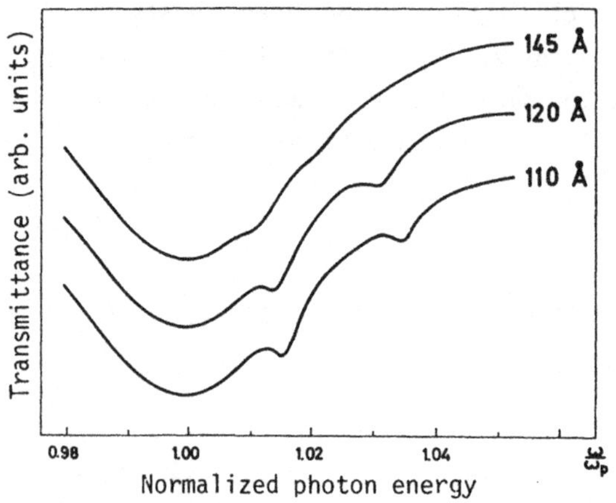

Fig. 3.8. It shows the transmittance of Ag films of different thickness for p polarised light. The angle of incidence is 75°. The minima indicate resonances for volume plasmon frequencies at the wave vectors $q \approx n\pi/D$ with $n = 3$ and 5 /3.19/.

4. The Energy Loss Spectrum of Electrons and the Loss Function

4.1 Evaluation of the Loss Spectrum

The evaluation of an electron loss spectrum allows one to derive the loss function $\mathrm{Im}(-1/\varepsilon)$ and applying the Kramers-Kronig relations /4.1/ the dielectric function $\varepsilon(\omega)$ of the transmitted substance. This interesting aspect needs a more detailed discussion of the procedure.

In an energy loss spectrum of electrons we measure the intensity of electrons (number per second): $I(\Delta E,\vec{\theta})$ having lost the energy ΔE and being scattered into the direction $\vec{\theta}$; $\vec{\theta}$ means a vector with $|\vec{\theta}| = \theta$ the scattering angle against the primary beam. The connection of the elementary probability, e.g., (3.3) with $I(\Delta E,\vec{\theta})$, is important for the quantitative verification of the physical assumptions of the dielectric theory; it shall be sketched briefly.

$$I(\Delta E,\vec{\theta}) = I_0(\Delta E,\vec{\theta}) \times \frac{\partial^3 P_{inel}}{\partial \Delta E \partial^2 \Omega} \tag{4.1}$$

with

$$I_0(\Delta E,\vec{\theta}) = I_{00}(\Delta E,\vec{\theta}) \times \frac{\partial^3 P_{el}}{\partial \Delta E \partial^2 \Omega} \tag{4.2}$$

$I_{00}(\Delta E,\vec{\theta})$: energy and angular distribution of the incoming electrons entering the film,

$$I_0(\Delta E,\vec{\theta})$$

the same distribution of the elastically scattered electrons behind the film,

$$\frac{\partial^3 P_{el,\ inel}}{\partial \Delta E \partial^2 \Omega}$$

elastic or inelastic scattering probability.

The differential inelastic probability can be written, since multiple scattering processes have to be regarded,

$$\frac{\partial^3 P_{inel}}{\partial \Delta E \partial^2 \Omega} = \sum_{\nu=1}^{n} \frac{\partial^3 P_\nu}{\partial \Delta E \partial^2 \Omega} \tag{4.3}$$

with the recursion relation

$$\frac{\partial^3 P_n}{\partial \Delta E \partial^2 \Omega} = \frac{1}{n} \int \frac{\partial^3 P_{n-1}(\Delta E'-\Delta E, \vec{\theta}'-\vec{\theta})}{\partial \Delta E' \partial^2 \Omega'} \quad \frac{\partial^3 W(\Delta E', \vec{\theta}')}{\partial \Delta E' \partial^2 \Omega'} \, d\Delta E' d^2\Omega' \tag{4.4}$$

and

$$\frac{\partial^3 P_o}{\partial \Delta E \partial^2 \Omega} = e^{-W} \cdot \delta(\Delta E, \vec{\theta}). \tag{4.5}$$

Equation (4.1) becomes then with (4.4)

$$I(\Delta E, \vec{\theta}) = \sum_{\nu} \int I_o(\Delta E'-\Delta E, \vec{\theta}'-\vec{\theta}) \frac{\partial^3 P_\nu}{\partial \Delta E' \partial^2 \Omega'} \, d\Delta E' d^2\Omega'. \tag{4.6}$$

Equation (4.6) shows how the inelastic intensity is composed of the multiple processes.

The elementary differential probability $\partial^3 W/\partial \Delta E \partial^2 \Omega$ is given in the case of volume plasmons by (3.3); in the case of excitation of surface plasmons (10.44) has to be taken.

For the further procedure we have to perform the energy- and angular convolution of (4.2). Since the energy distribution $f(\Delta E)$ and the angular distribution $g(\theta)$ of the elastically scattered electrons are independent on each other - this has been verified in the experiment - we can write (4.2) /4.2/

$$I_o(\Delta E, \vec{\theta}) = I_{oo}(0,0) \cdot f(\Delta E) \cdot g(\vec{\theta}) \tag{4.7}$$

with $g(0) = f(0) = 1$.

We can assume that these distributions are not changed by the passage of the electrons through the film; this can be controlled by the experiment. For example, in experiments with a Hartl monochromator (Fig. 11.4) and 50 keV electrons with an angular width of 0.23 mrad of the primary electrons in the arrangement of Fig. 11.1 an angular width of 0.25 mrad behind a Si film of 1000 Å thickness, of 0.27 mrad behind an Al film of 600 Å, and of 0.26 mrad behind a Ge film of 1300 Å thickness has been measured /4.1a/. This result allows one to identify the angular width behind a film with that before the film, if the thickness of the films has similar values or less.

If we neglect multiple processes, the sum (4.3) has only the term P_1 and we obtain the intensity of those electrons having lost one ΔE and being scattered into the direction $\vec{\theta}$ from (4.6) as

$$I_1(\Delta E,\vec{\theta}) = I_0(0,0)\cdot\int d\Delta E' d^2\Omega' \Big\{ f(\Delta E'-\Delta E)\cdot g(\vec{\theta}'-\vec{\theta}) \cdot \int e^{-W}\, \delta(\Delta E''-\Delta E', \vec{\theta}''-\vec{\theta}')\, \frac{\partial^3 W(\Delta E'',\vec{\theta}'')}{\partial\Delta E''\partial^2\Omega''}\, d\Delta E'' d^2\Omega'' \Big\} \quad (4.8)$$

$$= I_0(0.0)\cdot e^{-W}\cdot\int d\Delta E' d^2\Omega' \cdot f(\Delta E'-\Delta E)\cdot g(\vec{\theta}'-\vec{\theta})\cdot \frac{\partial^3 W(\Delta E',\vec{\theta}')}{\partial\Delta E'\partial^2\Omega'} \quad (4.9)$$

or with $W \ll 1$,

$$I_1(\Delta E,\vec{\theta}) = I_{00}(0,0)\,\frac{D}{(\pi a_0 e)^2 k_{el}^2}\int d\Delta E' \int d^2\Omega' \cdot \mathrm{Im}(-1/\varepsilon)\cdot \frac{f(\Delta E'-\Delta E)\; g(|\vec{\theta}'-\vec{\theta}|)}{\theta'^2-\theta_{\Delta E}^2} \quad (4.10)$$

with

$$d^2\Omega' = d\phi' \sin\theta' d\theta'. \quad (4.11)$$

The integration interval is given by the shape of the functions f and g around the loss ΔE. If $f(\Delta E)$ has an energy halfwidth of 0.5 - 0.7 eV, $\theta_{\Delta E}$ does not change its value in this interval and the integration over $d^2\Omega'$ can be separated from that over $\Delta E'$. We have then

$$\int \frac{g(|\vec{\theta}'-\vec{\theta}|)\, 2\pi \sin\theta' d\theta'}{\theta'^2-\theta_{\Delta E}^2}\cdot \int \mathrm{Im}(-1/\varepsilon)\cdot f(\Delta E'-\Delta E)\, d\Delta E'. \quad (4.12)$$

If the loss function has a larger width than the energy distribution $f(\Delta E)$ as is often the case, e.g., Si and Ge, we can regard $\mathrm{Im}(-1/\varepsilon)$ as nearly constant in this energy interval and obtain for (4.12)

$$\overline{\mathrm{Im}(-1/\varepsilon)}\cdot G\cdot E \quad (4.13)$$

with G and F the integrated functions f and g, see /1.1d, 4.2/. The loss function represents then an averaged value. Such a procedure is not allowed in Al with an energy halfwidth of about 500 meV. A deconvolution of $I_1(\Delta E,\vec{\theta})$ to derive the natural halfwidth of the plasmon loss with known functions $f(\Delta E)$ and $g(\vec{\theta})$ can be made in principle, but the procedure is up to now not practicable.

If the substances behave as a nearly free electron gas, the difficulty just mentioned can be overcome by using the dielectric function of a free electron gas and convoluting $\mathrm{Im}(-1/\varepsilon)$ with the measured functions $f(\Delta E)$ and $g(\vec{\theta})$. By changing the energy halfwidth a fit with the observed values can be made. Detailed measurements at Al, Si, and Ge demonstrated that the shape of the loss peak is well described by this procedure and yields the natural halfwidth. In Fig. 4.1a,b the observed loss peaks are compared with the calculated ones using the free electron model and the measured functions $f(\Delta E)$ and $g(\vec{\theta})$ at $\theta = 0$ and at $\theta = 0.5$ mrad. A good agreement is reached in both cases /4.3/.

The formula (4.10) opens two possibilities: the verification of the dielectric concept as well as the determination of the loss function and thus the dielectric function from the loss spectrum.

a) The dielectric theory can be verified. If one knows the values of $\varepsilon'(\omega)$ and $\varepsilon''(\omega)$ over a large frequency region from any optical measurements, e.g., from reflection experiments, the loss function dependent on the value of the energy loss $\Delta E = \hbar\omega$ can be calculated. Measuring further the angular and energy distribution of the no loss peak $g(\vec{\theta})$ and $f(\Delta E)$ the loss spectrum is obtained for the scattering angle $\theta \cong 0$ or $q = 0$, see (4.10). It can be compared with the absolute number of electrons per second per energy loss interval by measuring the film thickness D or by fitting the calculated curve with a known value of $\varepsilon'(\omega_1)$ at a given frequency ω_1 in general in the region of the visible light. Thus the comparison of the structure of the loss spectrum and its *absolute* value with the observed data is possible. Experiments of this type are in good agreement with the calculations /4.4/, see Fig. 5.1. For more details see /1.1d/.

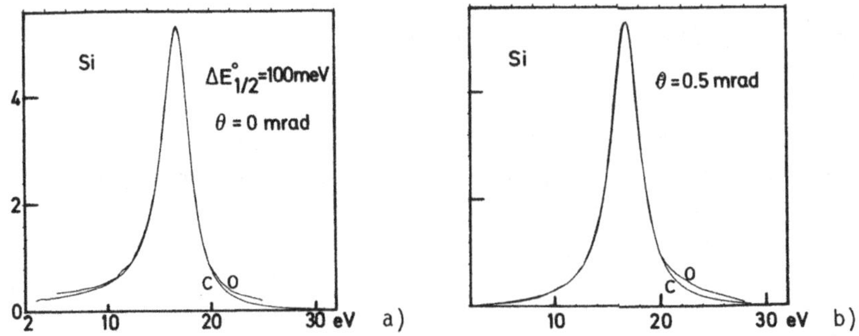

Fig. 4.1a,b. Volume plasmon of Si = 16.9 eV recorded at $\theta = 0$ (a) and $\theta = 0.5$ mrad (b). In addition a calculated curve (c) is drawn assuming a free electron gas behaviour ($\Delta E_{1/2} = 3.2$ eV), which is nearly identical with the observed curve (0) /4.3/

One has to keep in mind that this calculated loss intensity means the volume intensity at $\theta \cong 0$ and has to be completed by surface contributions. However the comparison of this (calculated) volume loss intensity can be made with a loss spectrum at larger scattering angles, see below, then the rapidly disappearing contributions of the surface can be neglected, see /1.1d/.

b) The loss function and the dielectric function can be derived from the loss spectrum. If the loss function ($\mathrm{Im}\{-1/\varepsilon\}$) is known over a large $\hbar\omega$ region, let us say between $\sim 1/2$ and ~ 50 eV, which corresponds to a wavelength region of some 10.000 Å up to some hundred Å, the application of the Kramers-Kronig relation allows one to determine the real part of $1/\varepsilon(\omega)$. From both functions $\mathrm{Im}(-1/\varepsilon)$ and $\mathrm{Re}(1/\varepsilon)$ the dielectric function $\varepsilon(\omega)$ can be determined or the refractive index n together with the absorption coefficient κ. The detailed procedure is described elsewhere /1.1d/.

This analysis has to be made at $q > 0$ ($\theta \sim 0.4$ mrad, 50 keV) to avoid the surface contributions. It is also possible to start with the analysis of the loss spectrum at $q \sim 0$ as shown in /4.5/. In addition to the corrections for surface contributions in general corrections for multiple scattering have to be made to obtain the loss function. This is also valid for finite q.

In most cases the condition $W \ll 1$ is only an approximation valid for rather thin films. In general multiple processes exist, which have to be considered if one derives $\varepsilon(\omega)$ from the loss spectrum /1.1d/. They produce higher loss intensities at larger ΔE values than are calculated with (4.10). If the film is not too thick, $D \lesssim \Lambda$, the correction for double losses has to be applied at least. The intensity of double losses has been calculated as /4.7/

$$I_2(\Delta E,\theta) = \int_0^D \int_0^{\Delta E} \int_0^{\theta_c} \frac{I_1(\Delta E_1,\vec{\theta}_1)}{D}\, dD_1 \cdot \frac{\partial^3 W(\Delta E - \Delta E_1, \vec{\theta}-\vec{\theta}_1)}{\partial \Delta E\, \partial^2 \Omega} \cdot \frac{D-D_1}{D}\, d\Delta E_1 d\vec{\Omega}_1 . \tag{4.13a}$$

Here D_1 means the length in the film after which the electron suffers the first loss ΔE_1 and the deviation $\vec{\theta}_1$ in the element dD_1. Having passed the length $D-D_1$ the electron suffers the second collision $(\Delta E_2,\vec{\theta}_2)$ with the probability

$$\frac{\partial^3 W}{\partial \Delta E\, \partial^2 \Omega}(\Delta E_2,\vec{\theta}_2) \cdot \frac{D-D_1}{D} . \tag{4.13b}$$

If we observe the loss $\Delta E = \Delta E_1 + \Delta E_2$ in the direction $\vec{\theta} = \vec{\theta}_1 + \vec{\theta}_2$, we obtain $I_2(\Delta E,\vec{\theta})$ by integration over dD_1, $d\Delta E_1$, and $d\Omega_1$. For those details see /4.6-8/.

The application gives satisfying results insofar as the corrections applied to the spectra of different film thicknesses lead to the same loss function in between the error of measurements, see e.g./4.7,24,25/.

4.2 Mean Free Path Length

Sometimes it is convenient to use the concept of the mean free path length (Λ) of an electron for one inelastic collision. To calculate this term we have to integrate the differential probability (3.3) over $d\Delta E$ and $d^2\Omega$ and we obtain the probability W of an electron of creating a volume plasmon $\hbar\omega_p$ in a free electron gas after having passed the thickness D and being scattered into all angles up to θ_c

$$W = \int d\Delta E \int d^2\Omega \, \frac{\partial^3 W}{\partial \Delta E \partial^2 \Omega} \ . \tag{4.14}$$

Using the relation valid for a free electron gas

$$\mathrm{Im}(-1/\varepsilon) = \frac{\pi}{2} \cdot \omega_p \cdot \delta(\omega - \omega_p) \tag{4.15}$$

and thus

$$\int_0^\infty \mathrm{Im}(-1/\varepsilon) \ \omega d\omega = \frac{\pi}{2} \cdot \omega_p^2 \tag{4.16}$$

we get the mean free path length Λ by setting W = 1 and D = Λ /4.9/

$$\Lambda = \frac{\hbar v^2}{e^2 \omega_p} \cdot \left(\ln \frac{v q_c}{\omega_p} \right)^{-1} \ . \tag{4.17}$$

This relation, valid in the relativistic region too, can be written for medium energies $v < c$ with $a_o = \hbar^2/me^2 = 0.5$ Å,

$$\Lambda = \frac{2 a_o E_o}{\Delta E} \ (\ln \theta_c/\theta_{\Delta E})^{-1} \ . \tag{4.18}$$

Regarding the dispersion of the volume plasmon the relation (4.17) is slightly changed into /4.10/

$$\Lambda = \frac{\hbar v^2}{e^2 \omega_p} \left(\ln \frac{v q_c}{1.13 \ \omega_p} \right)^{-1} \tag{4.19}$$

but the factor 1.13 has no large influence.

The mean free path length Λ increases for $v < c$ with the square of the electron velocity v and approaches nearly a constant at $v \sim c$; see also Fig. 4.2. The numerical value in Al ($\theta_c \cong 13$ mrad) for 50 keV electrons and $\Delta E = 15$ eV is $\Lambda = 800$ Å.

In order to measure Λ we need a nearly constant angular distribution of the primary electrons at least up to θ_c so that $g(\vec{\theta})$ can be regarded as constant and taken

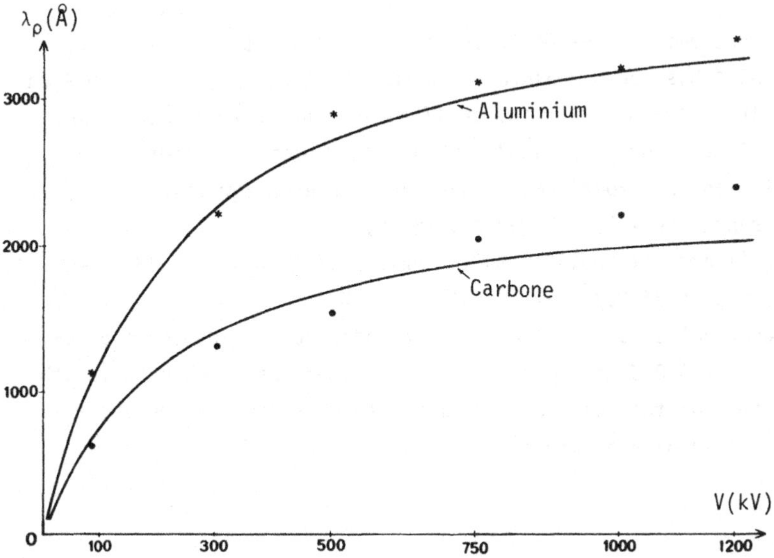

Fig. 4.2. Mean free path length for volume plasmons in Al and carbon as function of the electron energy; full line: calculated values (4.17); points: measured values /4.18/

out of the integral and an energy distribution $f(\Delta E)$ slowly varying compared with $\partial^3 W/\partial\Delta E \partial^2\Omega$; then we have from (4.9)

$$I_1/I_0 = D/\Lambda. \tag{4.20}$$

This quotient I_1/I_0 is then obtained experimentally by dividing the areas under the no loss peak and the one loss peak in the spectrum. With the known film thickness Λ can be determined.

In the case that the angular aperture of the primary electrons is smaller than θ_c corrections have been applied /1.1c,4.11,17/. In this way the following values of Λ of the Al 15 eV loss at E_0 = 50 keV have been obtained:

820 ± 150 Å	/4.11/	
720 ± 180 Å	/4.12/	360 ± 90 Å measured at 20 keV and extrapolated to 50 keV with (4.18)
650 ± 200 Å	/4.13/	interpolated
985 ± 200 Å	/4.14/	
695 ± 200 Å	/4.15/	
800 Å calculated		

Similar measurements were made on the Ge 16 eV loss /4.16/ and the Si 17 eV loss. For the Si loss Λ = 580 Å has been measured with 50 keV electrons, the calculated one is 750 Å /1.1c/. The value of Λ for the Be 17 eV loss has been found to be 270 ± 80 Å at 20 keV /4.12/. For evaporated carbon films which are important as substrates in transmission ELS experiments Λ has been measured for values of E_0 20 - 70 keV and different energy loss intervals /4.16,22/.

If the Λ values of Al are extrapolated to low values of E_0 using (4.18) they fit the results at 2 keV: ∼ 37 Å /4.23/, extrapolated ∼ 31 Å.

If one applies higher voltages up to 1200 keV one observes the dependence described by (4.17), as Fig. 4.2 demonstrates. The calculated mean free path length Λ (full line) as function of the voltage bends more and more down in agreement with the observed points in Al and carbon.

4.3 Multiple Losses

The experiments show that electrons having transmitted a thin foil of thickness D suffer not only one loss, but in general multiples of $\hbar\omega_p$, see Fig. 4.3. The higher

Fig. 4.3. Multiples of the 15 eV loss in the energy loss spectrum of Al at E_0 = 1200 keV /4.18/

losses ($n \cdot \Delta E$) are more intense when the foil is thicker. This result is described by the Poisson distribution which follows from integrating (4.4,5) which can be written as

$$P_n = \frac{1}{n!}\left(\frac{D}{\Lambda}\right)^n e^{-D/\Lambda} \ . \tag{4.21}$$

Figure 4.4 illustrates the dependence on D/Λ and demonstrates how the probability P_n for n losses increases with thickness passing a maximum at

$$(D/\Lambda)_{max} = n \ . \tag{4.22}$$

Whereas the no loss intensity P_0 decreases with D/Λ, $\sum_n P_n$ remains constant = 1. This has been verified in experiments on Al with 50 keV electrons, see Fig. 4.5 /4.19/. The same considerations hold for the multiples of surface plasmons /4.20/

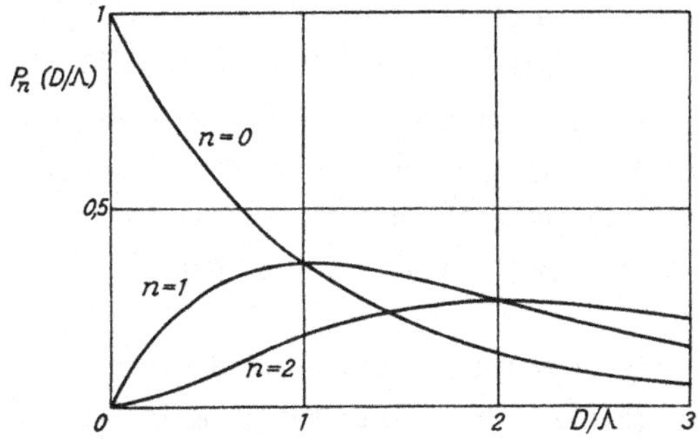

Fig. 4.4. The Poisson distribution $P_n(D/\Lambda)$ as function of D/Λ for n = 0, 1 and 2

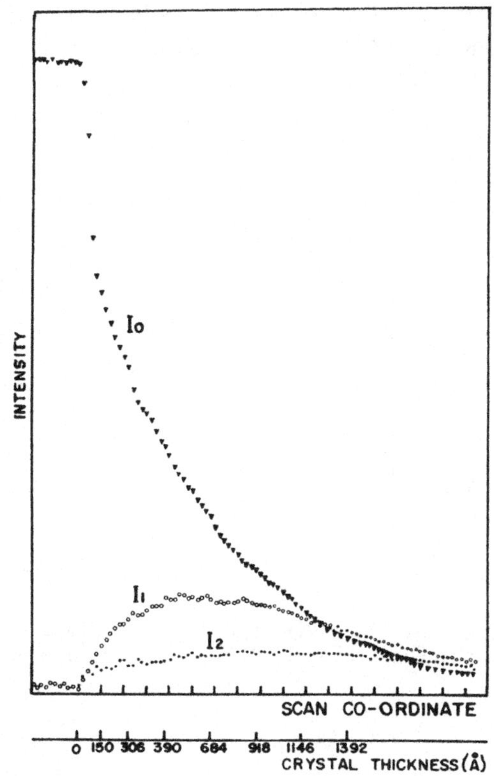

Fig. 4.5. Experimental elastic (I_0), first plasmon loss (I_1) and second loss (I_2) intensity as function of crystal thickness /4.19/

The multiple losses show two characteristic features:

1) The energy halfwidth of the multiple losses increases with the number n: The energy width of the n'th loss results from folding the energy distribution of the primary electrons with that of the loss peak given by the lifetime of the plasmon. Since this folding has to be repeated n times for the n'th loss the halfwidth of the energy distribution of the n'th loss increases with larger n, e.g. Ref. 4.11, Fig. 1.

2) The angular distribution of the intensity of multiple losses broadens at higher losses. This is evident since similar to the energy distribution multiple folding of the angular distribution broadens the resulting dependence as Fig. 4.6 shows for 50 keV electrons having passed a 5150 Å Al foil. The angular distribution of the primary beam had the value of $5.5 \cdot 10^{-4}$ rad, see also /4.17,21/. It follows that the intensity of higher losses persists at higher angles compared to the intensity of the lower losses. This is of interest for the evaluation of electron loss spectra especially if losses shall be separated from the background at $\theta > 0$.

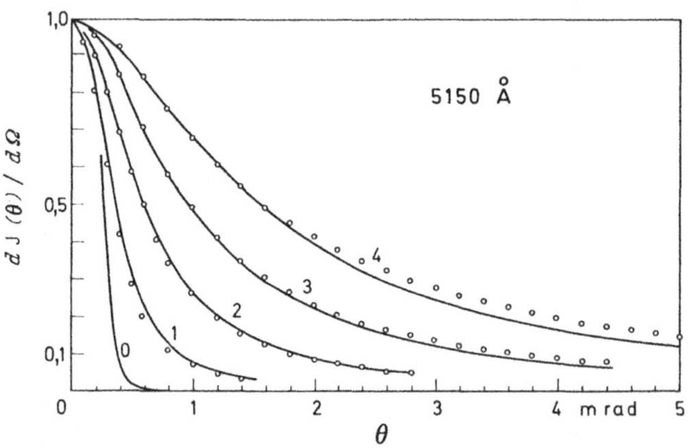

Fig. 4.6. Angular distribution of the Al 15 eV loss intensity and its multiples. Primary energy of the electrons 50 keV. Film thickness 5150 Å. The full lines are calculated values /4.21/

5. Experimental Results

In experiments with electrons in transmission of thin films or in reflection at clean surfaces a variety of loss spectra has been found. For a rough survey these can be described as follows:

a) The spectra of the substances with a nearly free electron gas (Al, Mg, Li, K, Na, etc.) dominated by a volume peak, and by a more or less strong surface loss peak if the transmitted film is thin enough.

b) The spectra of the wide gap insulators like those of the alkali halides, which consists of a number of peaks due to different interband transitions, see Fig. 5.1. Exciton peaks are visible too as is demonstrated by the example of solid neon, see Fig. 5.2.

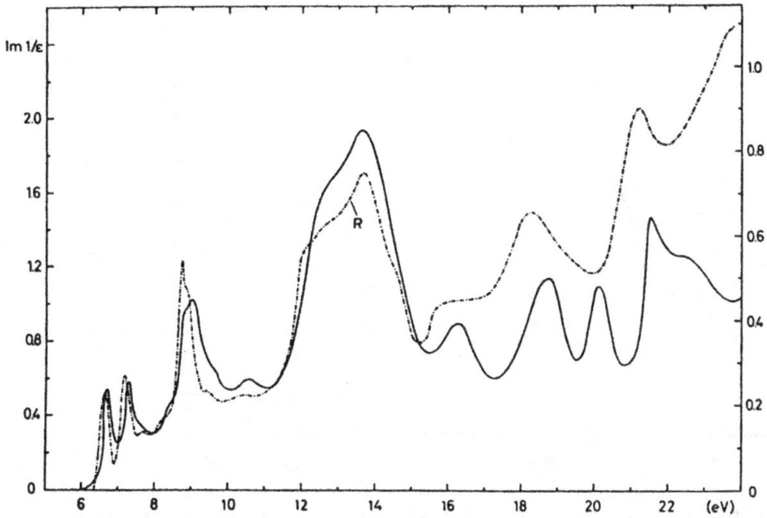

Fig. 5.1. Energy loss function of KBr deduced from electron loss experiments (full line) /5.1/ and loss function calculated from optical values (broken line) /5.2/ scale: right side /5.1/

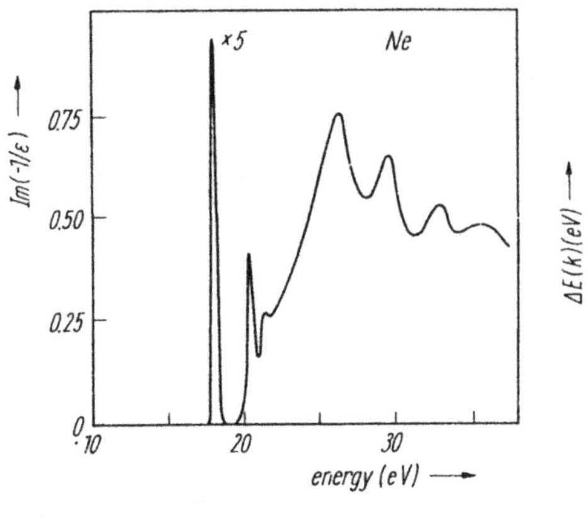

Fig. 5.2. Energy loss function of solid Ne which is dominated by the 17.6 eV exciton /5.3/

c) Insulators with small gap (semiconductors) have in general interband transitions at small energies of some eV and a pronounced volume peak due to the valence electrons, four electrons in the case of Si, Ge, III-V compounds. Its plasma energy is large compared with the gap so that the valence electrons can be regarded as nearly free. Beyond this plasmon peak we find transitions from deeper levels with characteristic structure, see the example of InSb, Fig. 5.3.

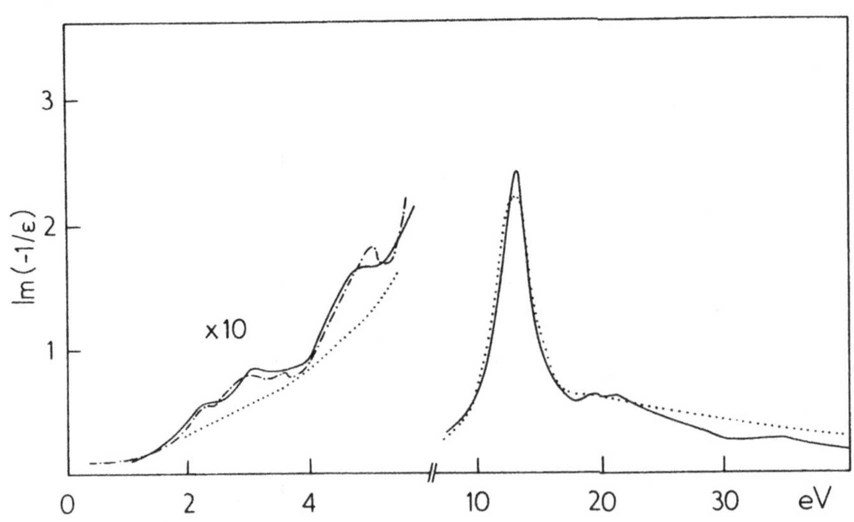

Fig. 5.3. Energy loss function of InSb. Besides the low-energy interband transitions the strong volume plasmon is observed. Full line: single crystalline /5.4/, dash-dotted: polycrystalline /5.5/, dotted line: amorphous InSb /5.4/

d) A large number of metals, e.g., the transition metals, have a loss spectrum with an entangled structure their details being mostly unexplained. Figure 5.4 shows an example of the loss function of gold, in addition the real and imaginary part of $\varepsilon(\omega)$, Fig. 5.5. One recognizes roughly the shape of the free electron gas function on which interband transitions are superposed.

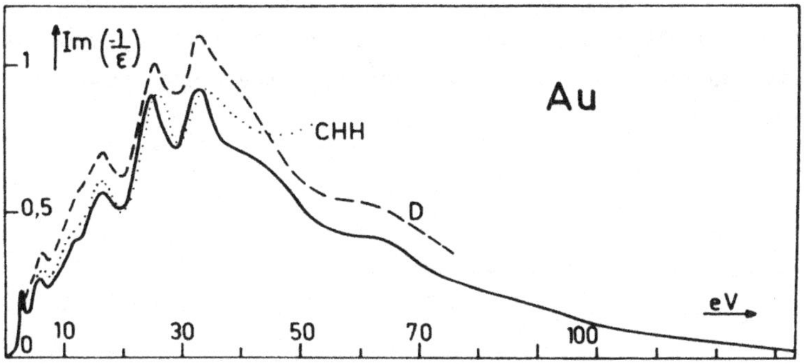

Fig. 5.4. Energy loss function of Au. Full line /5.6/, dashed line (D) /5.7/, dotted line: optical measurements CHH /5.8/

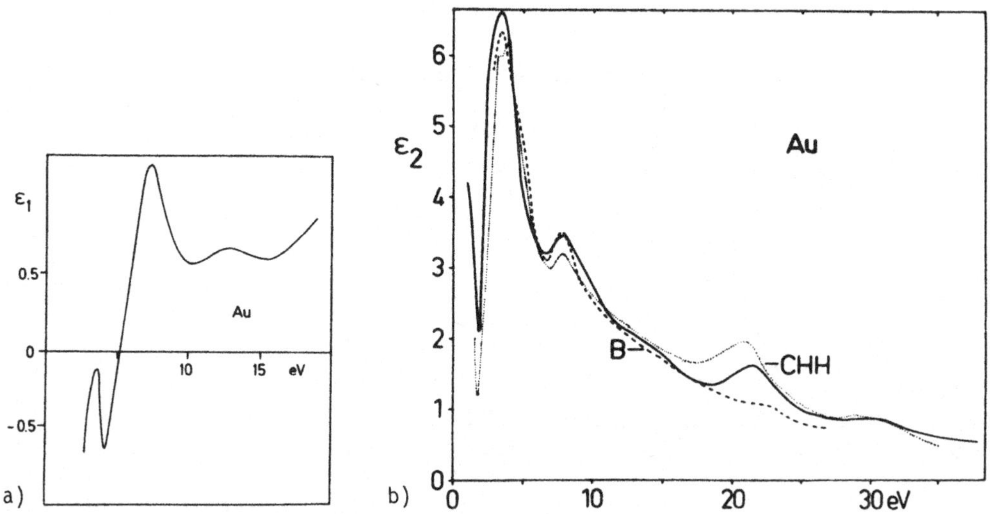

Fig. 5.5a,b. Dielectric function of gold. (a) Real part (ε_1), (b) imaginary part (ε_2) full line: from energy loss measurements /5.7/. Dashed /5.9/ and dotted /5.8/ line from optical measurements. This type of dielectric and loss function is characteristic for the noble and the transition metals, see /5.10/

e) In anisotropic crystals the spectrum depends on the direction of $\vec{q}$ inside the crystal, e.g., if in graphite the direction of q is oriented parallel to the c axis or to the a axis different loss functions and ε values are measured, see Figs. 5.6 and 5.7. The situation is more complex in biaxial crystals where the dielectric tensor has three different ε_{ii} values in the main axes. The imaginary part of the dielectric function of such a crystal is shown, e.g., for p-terphenyl in Fig. 5.8. For the evaluation of such spectra see /5.15/.

The loss intensity at low energies indicates whether we have to do with a metal or an insulator: In metals the loss function starts with zero at $\Delta E = 0$ and increases linearly with ΔE as can be calculated from the free electron loss function, whereas in substances with a band gap the loss function is zero up to the energy at which ΔE equals the gap energy.

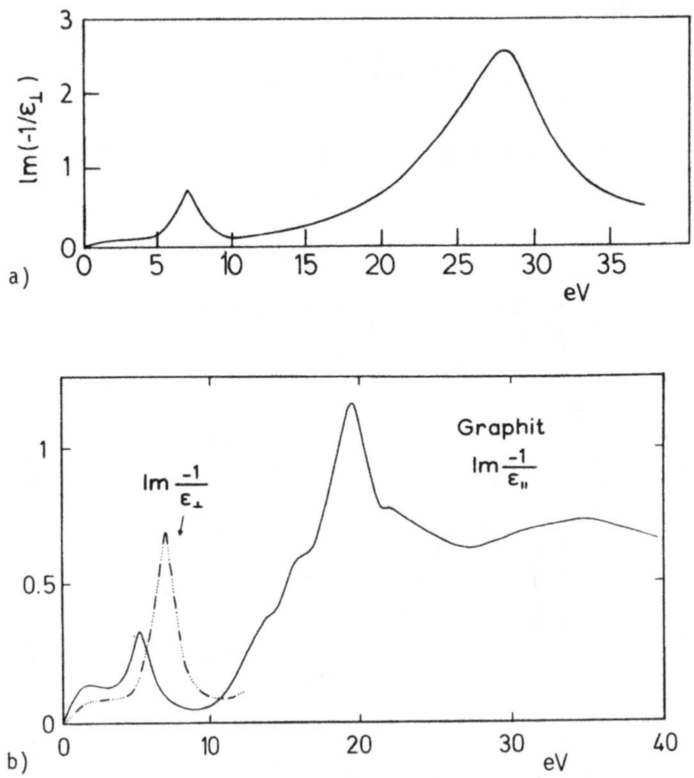

Fig. 5.6a,b. The loss function of an uniaxial crystal (graphite); (a) Im(-1/ε): $q_\perp$ c axis; (b) Im(-1/ε): $q_\parallel$ c axis. The peak at 7 eV is produced by the collective oscillation of the π electrons, that at 27 eV by the π plus σ electrons /5.11,12/

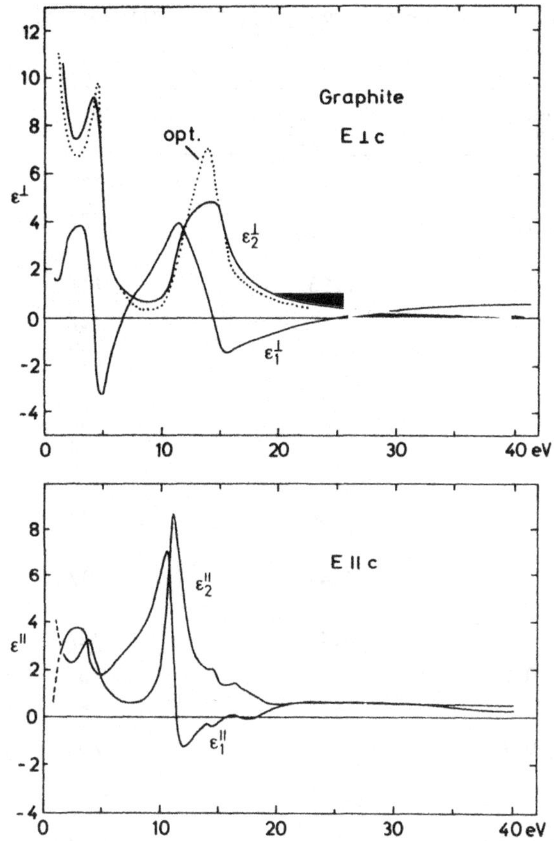

Fig. 5.7. The dielectric function $\varepsilon_{\perp}(\omega)$ and $\varepsilon_{\parallel}(\omega)$ of graphite deduced from the loss functions Figs. 5.6a,b /5.11,12/. For the electric wave vector perpendicular to the c axis optical data /5.13/ are added, dotted line

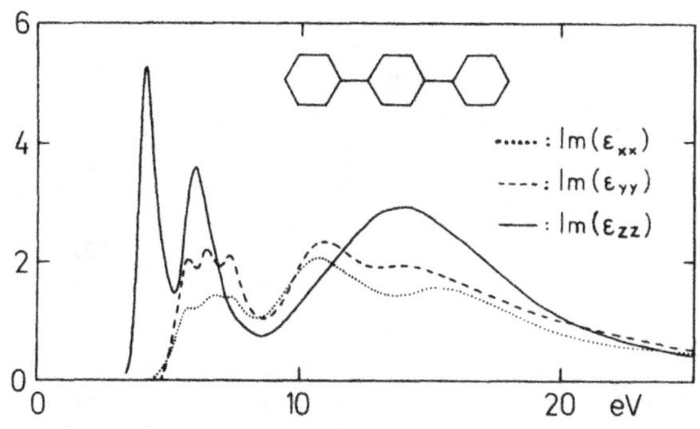

Fig. 5.8. The imaginary part of the dielectric function of monocrystalline p-terphenyl. The dielectric tensor of this biaxial crystal has three different values of ε_{ii} /5.14/

Table 5.1 contains a collection of experimental data of substances with a dominant volume plasmon; substances with less simple spectra, elements, and compounds are added in Table 5.2.

Table 5.1. Observed energy positions of the volume plasmons ΔE_p, the halfwidth $\Delta E_{1/2}$ and the theoretical value $\hbar\omega_p$

	ΔE_p, exp. [eV]	$\Delta E_{1/2}$, exp. [eV]	$\hbar\omega_p$, calc. [eV]	Ref.
Li	7.12	$\cong$ 2.5	8.02	/1/
	7.08 ± 0.05	2.2 ± 0.2		/2/
Na	5.72	$\cong$ 0.4	5.95	/1/
	5.70 ± 0.05	0.4		/2/
	5.6 ± 0.05	0.4		/3/
K	3.72	$\cong$ 0.4	4.29	/1/
	3.70 ± 0.05	0.3		/2/
Rb	3.41	$\cong$ 0.6	3.86	/1/
Cs	2.9	$\cong$ 1.2	3.44	/1/
Be	19		18.4	/4/
	18.5 ± 0.1	5.0 ± 0.2		/5/
	18.9 ± 0.1			/6/
	18.7 ± 0.2	4.4 ± 0.7		/3/
Mg	10.5		10.9	/7/
	10.39 ± 0.05	0.7 ± 0.1		/2/
	10.1 ± 0.1	0.6 ± 0.2		/3/
	10.35 ± 0.1	0.8		/8/
Ca	4.1			
	8.7		8.0	/2/
	4.3 ± 0.2	1.2 ± 0.2		
	8.8 ± 0.1	2.1 ± 0.2		/9/
Sr	4.0			/2/
	8.1		7.0	
	4.1 ± 0.2	1.5 ± 0.2		
	8.0 ± 0.1	2.3 ± 0.2		/9/
Ba	7.1		6.7	/2/
	7.2 ± 0.1	2.7 ± 0.2		/9/
Zn	17.8 ± 0.9			/10/
	17.0 ± 0.3			/6/

Table 5.1. (continued)

	ΔE_p, exp. [eV]	$\Delta E_{1/2}$, exp. [eV]	$\hbar\omega_p$, calc. [eV]	Ref.
Cd	19.2 ± 0.9			/10/
Hg	6.3 ± 0.1 solid	1.5	7.7	/11/
	6.4 ± 0.1 liquid	1.0	7.5	/11/
B	22.7 am.	∿ 18		/11a/
Al	14.97 ± 0.02	0.60 ± 0.02 Mo	15.8	/12/
	14.95 ± 0.05	0.5 ± 0.1		/2/
	14.92 ± 0.03	0.53 ± 0.05 Mo (crystal size > 250 Å)		/13/
	14.95 ± 0.06	0.52 ± 0.05 Mo		/14/
	15.03 ± 0.05	0.53 ± 0.05 Mo		/15/
	15.0 ± 0.2 LE			/16/
Ga	13.75 ± 0.05 cryst.	1.4	14.5	/17/
	13.70 ± 0.05 liq.	1.15		/17/
	13.84 ± 0.06 cryst.	0.6 Mo		/14/
	14.2 ± 1 LE			/18/
In	12.3		12.5	/19/
	11.4			/20/
	11.43 ± 0.1	0.39 ± 0.1		/21/
	11.4 ± 0.1	0.7 ± 0.1		/22/
Diamond	34 cryst.	14	31	/23,24/
	31 cryst. LE			/25/
Graphite	q⊥c 6.8 (π electron)		12.6	/26,27/
	27 (π + σ electrons)			
	q⊥c 7.2			/28/
	26.8			
Carbon	25.9	21.8 (evaporated film)		/28a/
Si	16.45 ± 0.1 cryst.	3.6	16.6	/29/
	16.1 ± 0.1 am.	4.0		/29/
	16.5 ± 0.1 cryst.	3.7 ± 0.2		/5/
	16.4 ± 0.1 am.	3.9 ± 0.2		/5/
	16.4 ± 0.3 am.	3.8 ± 0.5		/3/

Table 5.1. (continued)

	ΔE_p, exp. [eV]	$\Delta E_{1/2}$, exp [eV]	$\hbar\omega_p$, calc. [eV]	Ref.
Si	16.9 ± 0.1 cryst.	3.2 ± 0.2 Mo		/30/
	17.3 ± 0.4 cryst.	(111)face LE		/31/
Ge	16.5 cryst.		15.6	/5/
	15,90 am.		14.8	/5/
	16.49 cryst.			/32/
	15.90 am.			/32/
	15.95 ± 0.1 cryst.	3.4		/29/
	15.5 ± 0.1 am.	4.0		/29/
	16.0 ± 0.2 cryst.	3.3 ± 0.7		/3/
	15.8 ± 0.2 am.	4.1 ± 0.8		/3/
	15.9 ± 0.1 cryst.	3.4 ± 0.2		/5/
	15.5 ± 0.1 am.	3.6 ± 0.2		/5/
	16.2 ± 0.1 cryst.	3.1 ± 0.2 Mo		/30/
	16.4 ± 0.2 LE			/16/
Sn	13.67 ± 0.05	1.6	14.3	/17/
	13.40 ± 0.05 liq.	1.6		/17/
	13.7 ± 0.2	1.4 ± 0.2		/3/
	13.74 ± 0.06	1.20 ± 0.05 Mo		/14/
Ag	3.78 ± 0.02	0.075 ± 0.015 Mo		/51a/
Pb	13.25 ± 0.2		13.48	/6/
	12.96 ± 0.05			/33/
Sb	15.3	∿6	15.1	/34/
	15.2 ± 0.2	3.3 ± 0.5		/3/
	15.0 ± 0.5			/10/
	15.3 ± 0.1	3.3 ± 0.5		/35/
Bi	14.4 ± 0.2	6.8 ± 0.2	14.0	/36/
	14.1 ± 0.3	5.4 ± 1.5		/3/
	14.0 ± 0.3			/37/
Se	18.9 ± 0.2 cryst.	8.1 ± 0.4	17.4	/38/
	18.3 ± 0.2 am.	8.1 ± 0.4	16.4	/38/
Te	17.7 ± 1.1			/10/
	17.1 ± 0.1 cryst.	6.2 ± 0.3	15.6	/38/
	16.3 ± 0.1 am.	6.2 ± 0.3	15.3	/38/

Table 5.1. (continued)

	ΔE_p, exp. [eV]	$\Delta E_{1/2}$, exp [eV]	$\hbar\omega_p$, calc. [eV]	Ref.
As	18.66 cryst.		17.8	/32/
	16.69 am.		16.1	/32/
GaP	16.5 ± 0.2 cryst.		16.5	/39/
InAs	13.8 ± 0.2		13.8	/39/
GaSb	13.3 ± 0.2		13.8	/39/
GaAs	15.7 ± 0.2		15.7	/39/
	15.8 ± 0.1 cryst.			/41/
	15.5 ± 0.1 am.			/41/
InSb	12.8 ± 0.2		12.7	/39/
	13.0			/10/
	13.0 ± 0.2 cryst.			/20/
	13.2 ± 0.2 am.			/20/
	12.8 ± 0.1 cryst.			/41/
	13.1 ± 0.1 am.			/41/
SiC (α,β)	22.1	3.9 ± 0.3	22.9	/42/
PbS	15.3 ± 0.2		16.3	/43/
	14.5			/44/
PbSe	14.85 ± 0.2			/43/
	15.8			/45/
	15.1			/44/
PbTe	14.15 ± 0.2			/43/
	15.0			/45/
	14.1			/44/
AlAs	16.1		15.5	/140/
Al_2O_3	∿22	∿20		/44/

Table 5.2. Elements and compounds with less simple electron loss spectra listed together with the references

Solid Gases

O_2	/80/
N_2	/80/
Ne	/82/
Ar	/81/
Xe	/83/
CO_2	/80/
CH_4	/82/

Elements

Z		
21	Sc	/6,47,48/
22	Ti	/6,49,50,96(LE)/
23	V	/6,49,96(LE),97/
24	Cr	/6,49,96(LE),97/
25	Mn	/49/
26	Fe	/49,97/
27	Co	/49,97/
28	Ni	/49,97/
29	Cu	/16(LE),49,51,52,97,105,106/
41	Nb	/53/
42	Mo	/53,54,55(LE)/
46	Pd	/49,51/
47	Ag	/49,51,51a,56,57,106/
64	Gd	/47,58/
65	Tb	/47/
66	Dy	/47,58/
67	Ho	/47/
68	Er	/47/
69	Tm	/47/
70	Y	/6,47,48/
73	Ta	/53/
74	W	/53,59(LE)/
78	Pt	/49,51/
79	Au	/49,51,57,105/
80	Hg	/11/
92	U	/6,60/

Compounds

BeO	/61/
SiO_2	/62/
Mica	/62,63/
H_2O (solid)	/64/
As_2S_3	/65/
Mg_2Si	/66/
Mg_2Sn	/66/
Mg_2Ge	/66/
MgO	/67,141/
ZnO	/69,70(LE),71(LE),72(LE)/
MgF_2	/67,68/
CaF_2	/74/
SrF_2	/74/
BaF_2	/74/
BN	/75-77/
SiC	/77/
ReO_3	/78/
WO_3	/78,79/
$Na_{0.65}WO_3$	/78/
SnTe	/94/
CdO	/94/
CdTe	/10/
SmS	/137/
II/VI ZnS	/92/
ZnSe	/92/
ZnTe	/10,92/
CdS	/92/
CdSe	/92/
A15 V_3Si	/93(LE)/
Nb_3Al	/93(LE)/
Nb_3Sn	/93(LE)/
Nb_3Ge	/93(LE)/

Alkali halides /84-91/

TlCl	/95/
TlBr	/95/
TlI	/95/

Table 5.2. (continued)

Metaldichalcogenides		Alloys	
MoS_2	/28,73/	AgAu	/57,99/
$MoSe_2$	/28/	AgGa	/98/
WS_2	/28/	AlMg	/100,103/
WSe_2	/28/	AlCu	/100/
NbS_2	/28/	AlBe	/101/
$NbSe_2$	/28/	AlGe	/101/
TaS_2	/28/	AlSn	/101/
$TaSe_2$	/28/	AlZn	/104/
ZrS_2	/28/	InBi	/102/
$ZrSe_2$	/28/	InAl	/102/
HfS_2	/28/	CdMg	/103/
$HfSe_2$	/28/	CuNi	/103/
GaS	/28/		
GaSe	/28/		
SnS_2	/28/		
GeS	/46/		

Organic Substances

CH_4	Methane	/107/
C_5H_5N	Pyridine	/108/
C_6H_6	Benzene	/108-112/
$1,3C_6H_8$	1,3-Cyclohexadiene	/108/
$1,4C_6H_8$	1,4-Cyclohexadiene	/108/
C_6H_{10}	Cyclohexene	/108/
C_6H_{12}	Cyclohexane	/108/
C_6H_5Cl	Monochlorbenzene	/108/
C_7H_8	Toluene	/108/
$C_{10}H_8$	Naphtalene	/109,112-115/
$C_{12}H_{10}$	Diphenyl	/116/
$C_{13}H_{10}$	Fluoren	/117-119/
$C_{14}H_{10}$	Anthracene	/113,120-125/
$C_{14}H_{10}$	Phenanthrene	/113,121,127/
$C_{16}H_{10}$	Pyrene	/113,121,127-129/

Table 5.2. (continued)

Organic Substances		
$C_{18}H_{12}$	Chrysene	/113,121,127/
$C_{18}H_{12}$	Tetracene	/113,121,131/
$C_{18}H_{12}$	Triphenylene	/132/
$C_{18}H_{14}$	p-Terphenyl	/117,119,133,134/
$C_{20}H_{12}$	Benzpyrene	/121/
$C_{20}H_{12}$	Perylene	/131,132/
$C_{22}H_{12}$	Benzanthracene	/121/
$C_{22}H_{14}$	Picene	/121/
$C_{22}H_{14}$	Pentacene	/131,135/
$C_{24}H_{12}$	Coronene	/113,130,135/
Nucleic Acid Bases		
$C_5N_5H_5$	Adenine	/136/
$C_5N_2O_2H_6$	Thymine	/136/
$C_4N_2O_2H_4$	Uracil	/136/
PMMA	Polymethylmethacrylate	/138/
PS	Polystyrene	/139/
PVP	Poly(2-vinylpyridine)	/139/

References to Tables 5.1 and 5.2

1 C. Kunz: Z. Phys. 196, 311 (1966)
2 T. Kloos: Z. Phys. 265, 225 (1973)
3 O. Sueoka: J. Phys. Soc. Jpn. 20, 2203 (1965)
4 H. Watanabe: J. Phys. Soc. Jpn. 11, 112 (1956)
5 T. Aiyama, K. Yada: J. Phys. Soc. Jpn. 36, 1554 (1974)
6 E.A. Bakulin, L. Balabanova, M.M. Bredow, E.V. Stepin: Sov. Phys. Solid State 16, 1670 (1975)
7 P. Schmüser: Z. Phys. 180, 105 (1964)
8 C.H. Chen: J. Phys. Solid State Phys. 9, L321 (1976)
9 J. Langkowsky: J. Phys. D 8, 2058 (1975)
10 B. Gauthé, Phys. Rev. 114, 1265 (1959)
11 H. Boersch, J. Geiger, H. Hellwig, H. Michel: Z. Phys. 169, 252 (1962)
11a M. Urner-Wille: unpublished (1977)
12 J. Geiger, K. Wittmack: Z. Phys. 195, 44 (1966)
13 C. v. Festenberg: Z. Phys. 207, 47 (1967)

14 P. Zacharias: Z. Phys. 256, 92 (1972)
15 H.J. Hinz: Unpublished
16 J. Thirlwell: Proc. Phys. Soc. 91, 552 (1967)
17 K. Zeppenfeld: Z. Phys. 223, 32 (1969)
18 I.E. Rowe, I.C. Tracy, S.B. Christman: Surf. Sci. 52, 277 (1975)
19 J. Cazaux, C.R. Acad. Soc. Paris B 263, 244 (1969)
20 S. Zimmermann: J. Phys. C 9, 2643 (1976)
21 K.J. Krane: J. Phys. F 8, 2133 (1978)
22 T. Aiyama, K. Yada: J Phys. Soc. Jpn. 38, 1357 (1975)
23 K. Zeppenfeld: Thesis, Hamburg (1969)
24 J. Daniels, C. v. Festenberg, H. Raether, K. Zeppenfeld: "Optical Constants of Solids by Electron Spectroscopy", in *Springer Tracts in Modern Physics*, Vol. 54 (Springer, Berlin, Heidelberg, New York 1970) p. 78
25 N.R. Whetten: J. Appl. Phys. Lett. 8, 135 (1966)
26 K. Zeppenfeld: Z. Phys. 211, 391 (1968)
27 H. Venghaus: Phys. Status Solidi (b) 71, 609 (1975)
28 W.Y. Liang, S.L. Cundy: Philos. Mag. 19, 1031 (1969)
28a R.E. Burge, D.L. Misell: Philos. Mag. 18, 251 (1968)
29 K. Zeppenfeld, H. Raether: Z. Phys. 193, 471 (1966)
30 H.J. Hinz, H. Raether: Thin Solid Films 58, 281 (1979)
31 J.E. Rowe, G. Margavitondo, S.B. Christman, Phys. Rev. B. 15, 2195 (1977)
32 H. Richter, A. Rukwied: Z. Phys. 160, 473 (1960)
33 L.G. Bercht: Unpublished (1973)
34 J. Geiger: Z. Naturforsch. 17a, 696 (1962)
35 B. Bartning: Opt. Commun. 4, 404 (1972)
36 P. Zacharias: Opt. Commun. 8, 142 (1973)
37 J. Cazaux: C.R. Acad. Soc. Paris B 269, 1263 (1969)
38 N. Didden: Z. Phys. 257, 310 (1972)
39 C. v. Festenberg: Z. Phys. 227, 453 (1969)
40 R. Manzke: J. Phys. C 11, L349 (1978)
41 R. Manzke: In print
42 B. Bartning: Opt. Commun. 4, 78 (1971)
43 U. Büchner: Phys. Status Solidi (b) 83, 493 (1977)
44 L. Leder: Phys. Rev. 103, 1721 (1956)
45 B. Lahaye, F.H. Pradal, C. Gout: J. Phys. (Paris) 29, 137 (1968)
46 H. Venghaus, U. Büchner: Phys. Status Solidi (b) 72, 603 (1975)
47 C. Colliex, M. Gasgnier, P. Trebbia: J. Phys. 37, 397 (1976)
48 B. Brousseau-Lahaye, C. Colliex, J. Frandon, M. Gasgnier, P. Trebbia: Phys. Status Solidi (b) 69, 257 (1975)
49 C. Wehenkel, B. Gauthé: Phys. Status Solidi (b) 64, 515 (1974)
50 B. Brousseau, J. Frandon, A. Oustry, A. Escant: Surf. Sci. 71, 679 (1977)
51 J. Daniels: Z. Phys. 227, 234 (1969)
51a J. Daniels: Z. Phys. 203, 235 (1967)
52 T. Tagikawa, F. Fujimoto: J. Phys. Soc. Jpn. 40, 1630 (1976)
53 H.R. Apholte, K. Ulmer: Phys. Lett. 22, 552 (1966)
54 R. Manzke: In print
55 Y. Ballu, I. Lecante, H. Rousseau: Phys. Rev. B. 14, 3201 (1976)
56 A. Otto, E. Petri: Solid State Commun. 20, 823 (1976)
57 M. Schlüter: Z. Phys. 250, 87 (1972) (and alloys Ag + Au)
58 J. Daniels: Opt. Commun. 3, 13 (1971)
59 P.E. Luscher: Surf. Sci. 66, 167 (1977)
60 C. Wehenkel, B. Gauthé: Opt. Commun. 11, 64 (1974)
61 R. Gründler, K. Breuer, W. Tews: Phys. Status Solidi (b) 86, 329 (1978)
62 U. Büchner: J. Phys. C 8, 2781 (1975)
63 A.I. Atkins, D.L. Misell: J. Phys. C 5, 3153 (1972)
64 J. Daniels: Opt. Commun. 3, 240 (1971)
65 J. Perrin, J. Cazaux, P. Soukiassian: Phys. Status Solidi (b) 62, 343 (1974)
66 J. Tejeda, M. Cardona, N.I. Shevchik, D.W. Langer, E. Schönherr: Phys. Status Solidi (b) 58, 189 (1973)
67 H. Venghaus: Opt. Commun. 2, 447 (1971)
68 C. Gout, B. Lahaye, P. Perrier: Compt. Rend. (Paris) B 265, 1460 (1967)

69 I.I. Hengehold, R.I. Almassy, F.L. Pedrotti: Phys. Rev. B 1, 4784 (1970)
70 H. Froitzheim, H. Ibach: Z. Phys. 269, 17 (1974)
71 Y. Margoninsky, D. Eger: Phys. Lett. 59A, 305 (1976)
72 R. Dorn, H. Lüth: J. Appl. Phys. 47, 5097 (1976)
73 K. Zeppenfeld: Opt. Commun. 1, 377 (1970)
74 J. Frandon, B. Lahaye, F. Pradal: Phys. Status Solidi (b) 53, 565 (1972)
75 R. Vilanove: C.R. Acad. Sci. Paris 272, 1066 (1971)
76 U. Büchner: Phys. Status Solidi (b) 81, 227 (1977)
77 J. Cazaux: Compt. Rend. (Paris) B 270, 700 (1970)
78 R.E. Dietz, M. Campagna, N. Chazaviel, H.R. Shanks: Phys. Rev. B 17, 3790 (1978)
79 J.J. Ritsko, H. Witzke, S.K. Deb: Solid State Commun. 22, 455 (1977)
80 J. Daniels: Opt. Commun. 2, 352 (1970)
81 O. Bostanjoglio, L. Schmidt: Phys. Lett. 22, 130 (1966)
82 J. Daniels, P. Krüger: Phys. Status Solidi (b) 43, 659 (1971)
83 P. Keil: Z. Phys. 214, 251 (1968)
84 M. Creuzburg: Z. Phys. 196, 433 (1966)
85 M. Creuzburg, H. Raether: Solid State Commun. 2, 345 (1964)
86 F. H. Pradal, C. Gout, B. Lahaye: Solid State Commun. 4, 119 (1966)
87 O. Sueoka: J. Phys. Soc. Jpn. 19, 2239 (1964)
88 O. Sueoka: J. Phys. Soc. Jpn. 20, 2236 (1965)
89 P. Keil: Z. Phys. 214, 266 (1968)
90 C. Gout, F.H. Pradal: J. Phys. Chem. Solids 29, 581 (1968)
91 J. Maruyama, R. Onaka: J. Phys. Soc. Jpn. 44, 196 (1978)
92 T. Tomota, M. Mannami: J. Phys. Soc. Jpn. 27, 1204 (1969)
93 L.Y. Shen: Surf. Sci. 60, 239 (1976)
94 G. Schäfer: Z. Angew. Phys. 30, 345 (1971)
95 T. Sakurai, M. Mannami: J. Phys. Soc. Jpn. 27, 1703 (1969)
96 G. Simmons, E.J. Scheibner: J. Appl. Phys. 43, 693 (1972)
97 D.L. Misell, A.J. Atkins: Philos. Mag. 27, 95 (1973)
98 B. Bernert, P. Zacharias: Z. Phys. 241, 205 (1971)
99 O. Sueoka, F. Fujimoto: J. Phys. Soc. Jpn. 20, 569 (1965)
100 C.J. Powell: Aust. J. Phys. 13, 145 (1960)
101 O. Sueoka: J. Phys. Soc. Jpn. 20, 2212 (1965)
102 C.J. Powell: Philos. Mag. Suppl. 16, 203 (1967)
103 B.M. Hartley, J.B. Swan: Aust. J. Phys. 23, 655 (1970)
104 R.F. Cook, S.L. Cundy: Philos. Mag. 20, 665 (1969)
105 L.A. Feldkamp, L.C. Davis, M.B. Stearns: Phys. Rev. B 15, 5535 (1977)
106 H. Raether: "Solid State Excitations by Electrons (Plasma Oscillations and Single Electron Transitions)", in *Springer Tracts in Modern Physics*, Vol. 38 (Springer, Berlin, Heidelberg, New York 1965) p. 84
107 J. Daniels, P. Krüger: Phys. Status Solidi (b) 48, 659 (1971)
108 U. Killat: J. Phys. C 7, 2396 (1974)
109 R.N. Compton, R.H. Huebner, P.W. Reinhardt, L.G. Christophorou: J. Chem. Phys. 48, 901 (1968)
110 E.N. Lassettre, A. Skerbele, M.A. Dillon, K.J. Ross: J. Chem. Phys. 48, 5066 (1968)
111 U. Killat: Z. Phys. 263, 83 (1973)
112 E.E. Koch. A. Otto: Chem. Phys. Lett. 12, 476 (1972)
113 E.E. Koch, A. Otto: Opt. Commun. 1, 47 (1969)
114 E.E. Koch, A. Otto, K. Radler: Chem. Phys. Lett. 16, 131 (1972)
115 R.H. Huebner, S.R. Mielczarek, C.E. Kuyatt: Chem. Phys. Lett. 16, 464 (1972)
116 M. Shiko: J. Phys. Soc. Jpn. 36, 1636 (1974)
117 H. Venghaus: Thesis, Hamburg (1973)
118 H. Venghaus, H.J. Hinz: Phys. Status Solidi (b) 65, 239 (1974)
119 H.J. Hinz, H. Venghaus: Proc. IV Int. Conf. Vac. UV Rad. Phys., Hamburg (1974), p. 359
120 H. Venghaus: Z. Phys. 239, 289 (1970)
121 G. Jäger: Ann. Phys. 22, 147 (1969)
122 E.E. Koch, A. Otto, S. Kunstreich: Opt. Commun. 2, 365 (1971)
123 E.E. Koch, A. Otto, K. Radler: Chem. Phys. Lett. 21, 501 (1973)
124 S. Kunstreich, A. Otto: Chem. Phys. 3, 370 (1974)

125 S. Kunstreich, A. Otto: Chem. Phys. 3, 384 (1974)
126 E.E. Koch, A. Otto: Phys. Status Solidi (b) 51, 69 (1972)
127 H. Venghaus: J. Chem. Phys. 57, 3478 (1972)
128 H. Venghaus: Phys. Status Solidi (b) 54, 671 (1972)
129 H.J. Hinz, H. Venghaus, J. Chem. Phys. 64, 30 (1976)
130 L.Ya. Bubnov, E.L. Frankenich: Sov. Phys. Solid State 12, 1726 (1971)
131 Y. Kamula, K. Seki, H. Ionokuchi: Chem. Phys. Lett. 30, 35 (1975)
132 E.E. Koch, A. Otto: Chem. Phys. Lett. 6, 15 (1970)
133 H.J. Hinz, H. Venghaus: J. Chem. Phys. 62, 4937 (1975)
134 A. Hermann: Appl. Phys. 15, 185 (1978)
135 N. Ueno, Y. Hayiasi, S. Kiyono: J. Phys. Soc. Jpn. 44, 282 (1978)
136 M. Isaacson: J. Chem. Phys. 56, 1803 (1972)
137 J.E. Rowe, M. Campagna, S.B. Christman: Phys. Rev. Lett. 36, 148 (1976)
138 J.J. Ritsko, L.Z. Brillson, R.W. Bigelow, T.J. Fabish: J. Chem. Phys. 69, 3931 (1978)
139 J.J.Ritsko, R.W. Bigelow: J. Chem. Phys. 69, 4162 (1978)
140 R. Ludeke, L. Esaki: Surf. Sci. 47, 132 (1975)
141 V.E. Henrich, G. Dresselhaus: Phys. Rev. Lett. 36, 158 (1976)

Remarks to Tables 5.1 and 5.2

a) General. In Table 5.1 the measured volume loss positions (ΔE_p) and their half-width ($\Delta E_{1/2}$) are collected, whereas in the third column the calculated volume losses ($\hbar\omega_p$), given by $\omega_p = ne^2/m\tilde{\varepsilon}_0$ (1.8) are listed. For surface plasmons see Table 10.3.

LE means that the data are obtained from a reflection loss spectrum of low-energy electrons.

Mo indicates that a monochromator has been used (frequent energy width of the primary electrons 80-120 meV) so that $\Delta E_{1/2}$ is near to the natural line width. In the measurements of Al, Si, and Ge /15,30/ in Tab. 5.1 the free electron model has been used, so that the $\Delta E_{1/2}$ data can be regarded as the natural line width, see Chap. 4.1.

The specimens under investigation are in general polycrystalline, sometimes monocrystalline or at least epitaxially grown which is indicated by "cryst."; but liquid and amorphous (am.) phases have also been measured.

b) Influence of density. As the plasma frequency depends on the electron density n, the peak position of the volume loss (ΔE_p) changes with the temperature and can be different in the crystalline and amorphous phases and in the liquid. Since the density increases in general with decreasing temperature, ΔE_p displaces to higher energies if the crystal is cooled. Measurements with electrons of about 50 keV in transmission of thin foils were made between 700 and 295 K on Al /5.16, 17/ and between 295 and 10 K on Al and Pb /5.17/. Table 5.3 shows the result of the measurements at low temperatures: the peak position shifts by about 100 meV to higher values if one lowers the temperature from 295 to 10 K, in agreement with the calculated values.

Table 5.3. Loss energy difference between 10 and 295 K

	Al	Pb
obs. $\delta(\Delta E_p)$	105 ± 20 meV	150 ± 50 meV
calc. $\delta(\Delta E_p)$	90 meV	135 meV

In Table 5.3 $\delta(\Delta E_p)$ means the shift of $\{\Delta E_p\ (10\ K) - \Delta E_p\ (295\ K)\}$

It may be mentioned that the loss spectrum of Pb does not change when the metal reaches the supraconducting state /5.17/.

Some substances change the density from the solid into the liquid state by a certain amount, e.g., the solid state of In has a density of 7.28 at 20 ^{0}C whereas the liquid at 300 ^{0}C has a density of 6.91. Measurements of the surface loss at these temperatures showed a difference of 8.8 eV - 8.4 eV = 0.3 eV; the calculated value is 0.22 eV /5.18/. Similar changes are observed in Al /5.19/.

In certain substances the loss width of the liquid state has a smaller value than that of the solid state; examples are Hg and Ga, see Fig. 5.9.

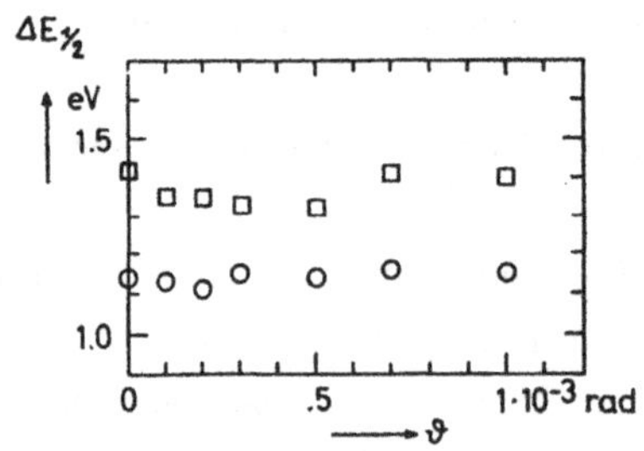

Fig. 5.9. The halfwidth of the plasmon loss of solid (□) and liquid (o) Ga at small q values. (The $\Delta E_{1/2}$ values are measured with an energy resolution of ∿ 0.7 eV and have not been corrected, so that the data of the figure are larger than the natural linewidth) /5.20/

The substances listed in Table 5.1 have been measured partially in the amorphous or in the liquid state. The amorphous solids being in general less dense than the crystalline one, their volume peak displaces to lower energies; an exception is InSb, see Table 5.1. This indicates that the density effect alone is not sufficient to explain the displacement; certainly the change of the band structure which parallels the phase change is of influence. The latter will also influence the loss width, since the dielectric function $\varepsilon_{GG'}(\omega,q)$ depends on the band structure. Up to now calculations have not yet been done to prove this possibility in a special case.

c) Comment to interband transitions. In addition to the plasmon peaks in Table 5.1 transitions with varying intensity are found which are associated with the band structure.

The interpretation of the band transitions in the loss spectra is done in most cases more or less qualitatively; sometimes special transitions are discussed. A few examples are given:

The III-V compounds have small interband maxima at the low-energy side of ΔE_p, see Fig. 5.3, which have been associated with the band structure. An interesting feature is the existence of losses beyond ΔE_p which are due to transitions from d core levels of the metal atom /5.10/; they split up by spin orbit coupling as optical observations showed /5.21/.

The alkali halides are a group of substances which have been studied in several papers, see Table 5.2. Besides the excitons the rather broad and dominant peak is often interpreted as volume plasmon. It is surprising that the observed ΔE values are so near to $\hbar\omega_p$ calculated with 6 or 8 electrons per molecule, although many neighbouring transitions are present, Table 5.4. Probably they compensate each other. In every case the shape of the $\varepsilon'(\omega)$ and $\varepsilon''(\omega)$ curves of the alkali halides are similar to those of a free electron gas in the region of its volume peak /5.22/: increasing ε' values starting from rather low values and decreasing ε'' values to smaller damping, as in Fig. 1.1, so that a polarisation wave can travel through the crystal.

Table 5.4. ΔE values of the dominant losses in some alkali halides /5.23/

	LiF	NaCl	KBr	RbI
obs.	25.3 eV	15.5 eV	13.2 eV	11.1 eV
calc.				
6 El/Mol	22.7 eV	13.6 eV	10.7 eV	9.2 eV
8 El/Mol	26.2 eV	15.7 eV	12.4 eV	10.6 eV

Looking at these losses at low temperatures a fine structure becomes visible in KBr as Fig. 5.10 demonstrates /5.1/. Here besides the excitons a, b, c and the excitons from deep levels (g), which display the spin orbit splitting of the initial level, the fine structure of the main loss (d) becomes more pronounced with lower temperatures. It is still open how this is compatible with the plasmon character of this loss.

As we shall see, Chap. 7.1, these losses, e.g., the LiF 25.3 eV show practically no dispersion, a behaviour different from that of a normal volume plasmon. A discussion of the origin of the losses in LiF between 10 and 70 eV is to be found in /5.1,24,25/.

d) There exist many data on loss spectra of *organic* solids. The aim of these experiments is to compare the observed loss spectra of the single molecules with the spectra of the crystal. By comparison information is expected as to which degree

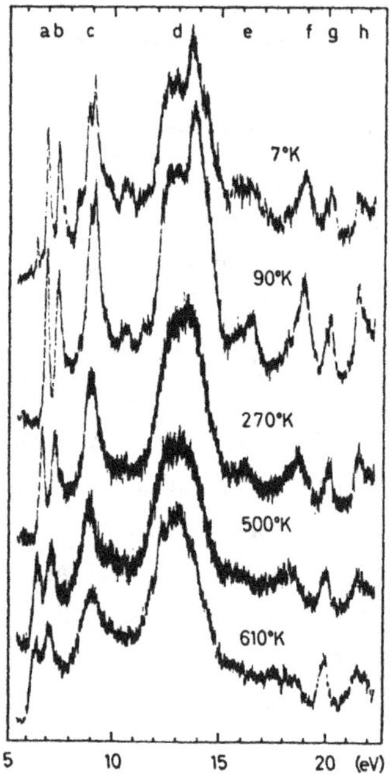

Fig. 5.10. Influence of temperature on the loss spectrum of KBr /5.1/

the molecular excitations persist in the solid. There exists a certain similarity in the π - π^* excitations in case of aromatics between molecules and solid, but at higher energies this is lost.

As far as one can see today the technical development for the investigation of direct vertical band transitions is better developed in optics, but there is no doubt that ELS can be improved.

e) Graphite intercalation compounds. As is known from chemical studies a variety of materials, for example, alkali metals, $FeCl_3$, etc., can be intercalated into the graphite planes. In the case of $FeCl_3$ intercalation the C-C interplanar distance changes from 3.35 to 9.38 Å. Measurements of reflectivity revealed a plasma edge in the energy range of about 1 - 3 eV due to an electron gas produced by the intercalated material. ELS of $FeCl_3$ intercalated graphite showed that the loss spectrum of graphite is only weakly perturbated insofar as the original 7 eV π plasmon peak is displaced to 5.9 eV. In addition a characteristic strong loss at 1.1 eV (halfwidth 0.3 eV) appears which supports the interpretation of the optical measurements. Detailed experiments with this 1.1 eV demonstrated a linear dependence of its energy position on q^2 nearly up to 10 mrad (E_0 = 80 keV) and a similar dependence of the loss width as has been found at volume plasmons too. A difference consists in the intensity dependence on q which has been found as $\sim q^{-n}$ with $n \cong 4$ whereas the volume plasmons follow n = 2. Such a stronger decrease of the intensity with $n \cong 3$ has been measured also on Ag (ΔE = 3.8 eV) and is explained as a damping effect by interband transitions (nonvertical direct transitions) coming into play at larger q values /5.30/. A number of interband transitions are observed in the loss spectrum which supports this explanation /5.25a/.

Radiation damage. It is well known that fast electrons produce radiation damages in crystals. Such changes have also been observed with ELS.

1) Irradiation of certain alkali halides produces additional loss peaks which can be attributed to alkali metals /5.26/. For example, NaBr films, made by vaporization, about 1000 Å and more thick, show a loss spectrum with two peaks, a narrow one at 5.7 eV and a larger one at 2.2 eV, Fig. 5.11. These additional losses which lie in the gap of NaBr or below 6.6 eV and are thus well separated from the

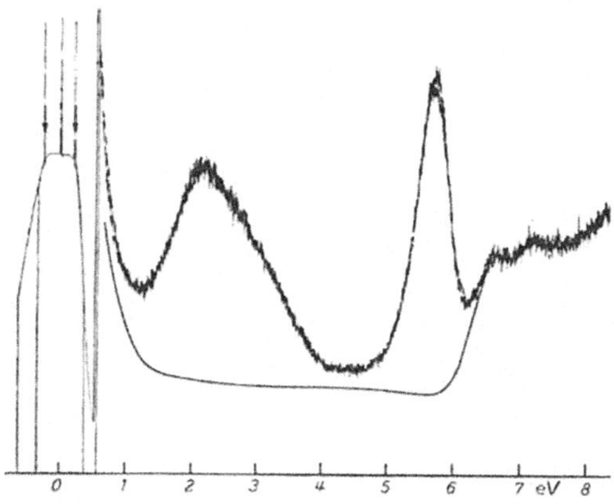

Fig. 5.11. Section of the NaBr loss spectrum after irradiation with electrons. The two strong losses are the plasmons of Na /5.26/

NaBr loss spectrum, become detectable at an irradiation with an intensity of 2.5 10^{-5} A/cm^2 after about 10 min. Similar results are obtained at NaI: losses at 5.7 and 2.3 eV appear after irradiation. Their intensity increases up to a maximum and remains then constant. The loss at 5.7 eV agrees with the volume loss of Na, whereas the lower loss can be regarded as the surface plasmon of Na (3.85 eV), displaced to lower values; such a lowering of the plasma frequency is due to the dielectric constant of the alkali halide in which the metal is embedded, see (10.9b) and probably to the geometrical shape of the metal. In case of a sphere the surface plasmon energy decreases from $\omega_p/\sqrt{2}$ to $\omega_p/\sqrt{3}$, see (10.64). Both effects could not be separated. The metal has been identified by its electron diffraction pattern. Further results were reported on KF, KI, and RbF.

2) Another type of damage has been reported on the alkali halides LiF, LiCl, and NaCl, which have been irradiated with X-rays before being measured by ELS. Several additional peaks are registered; a part of them can be associated with F centers known from optical measurements /5.27/.

3) In general the substance show no changes of the loss spectrum during the measurement. This is also valid for the organic substances listed in Table 5.3. The spectrum of 1.3 cyclohexadiene, recorded after 30 min irradiation with a current density of 10^{-4} A/cm^2 - this means a dose of 0.2 coul/cm^2 - however, shows a strong change which announces a decomposition of the substance /5.28/. Similar decomposition is reported on cholesterol /5.29/.

6. The Loss Width

In the following two chapters the energy position of the plasmons and their halfwidth are discussed as functions of the wave vector q. As already mentioned electrons which are inelastically scattered by an angle θ transfer a momentum $\hbar q = \hbar k_{el}\theta$ to the excited electrons (we assume here $\hbar q_{\perp} > \hbar q_{\parallel}$). Measurements of the peak position and the halfwidth as function of θ allow one to derive their q dependence.

6.1 The Dispersion of the Loss Width

Observations of the loss spectra at different q values demonstrated that the halfwidth of the volume losses ($\Delta E_{1/2}$) increases with larger q values /6.1,2/. Figure 6.1 shows the halfwidth of some substances as function of θ^2. In the first approximation the experimental dependence of $\Delta E_{1/2}(\theta)$ can be described by the relation

$$\Delta E_{1/2}(\theta) = E_{1/2}(0) + B\theta^2. \tag{6.1}$$

Table 6.1. Values of the dispersion coefficient B (10^4 eV/rad)

Al	∿ 1.4	/6.3/
	1.93 ± 0.2	/6.4/
In	3.58 ± 0.3	/6.4/
Mg	1.5	/6.3/
Li	∿ 4.7 for $q > 0.5$ Å^{-1}	/6.3/
Na	∿ 3.4	/6.3/
K	∿ 3.5	/6.3/
Sn	∿ 0.15	/6.6/

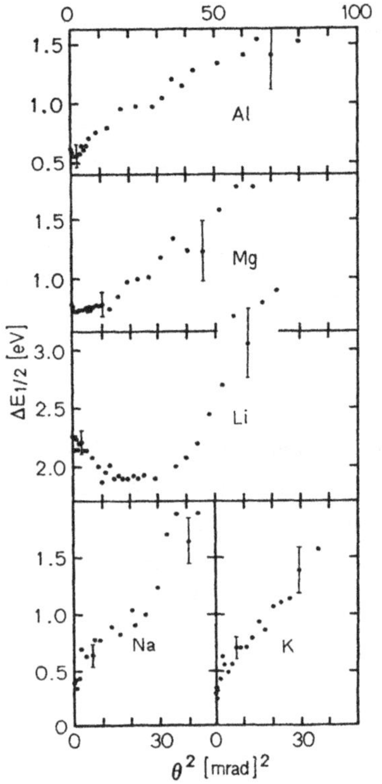

Fig. 6.1. Dependence of the halfwidth $\Delta E_{1/2}$ of different volume plasmons on the square of the scattering angle θ. E_0 = 50 keV, k_{el} = 117 Å^{-1} /6.3/

Values of /6.5/ have not been included in Table 6.1, since they differ rather strongly from these data. Polycrystalline and liquid Sn show a different dependence $\Delta E_{1/2}(\theta)$ for both phases: B (liquid) is nearly 2 times B (cryst.). Both curves seem to approach at higher q values within the limits of error /6.6/. A difference of the halfwidths at small q values has been observed in Ga too, see Fig. 5.9.

Experiments with X-rays confirmed these results of the increase of the linewidth as function of q, which has been observed at the 19.4 eV Be loss /6.7/. It is interesting that measurements on a Be-monocrystal (hexagonal) showed that this dependence has an anisotropic character: if q is parallel to the a axis the width and its increase are larger than in the direction of the c axis /6.8/, see Fig. 6.2.

A similar anisotropy of the $\Delta E_{1/2}$ (q) dependence is reported on single crystals of Mg in experiments with fast electrons /6.9/; in contrast to Be the plasmon width along the c axis is broader in Mg than along the a axis. An interesting observation has been made in Li /6.3/: here the value of $\Delta E_{1/2}$ goes through a minimum at $q \cong 0.5$ Å^{-1} beyond this value $\Delta E_{1/2}$ increases with q^2.

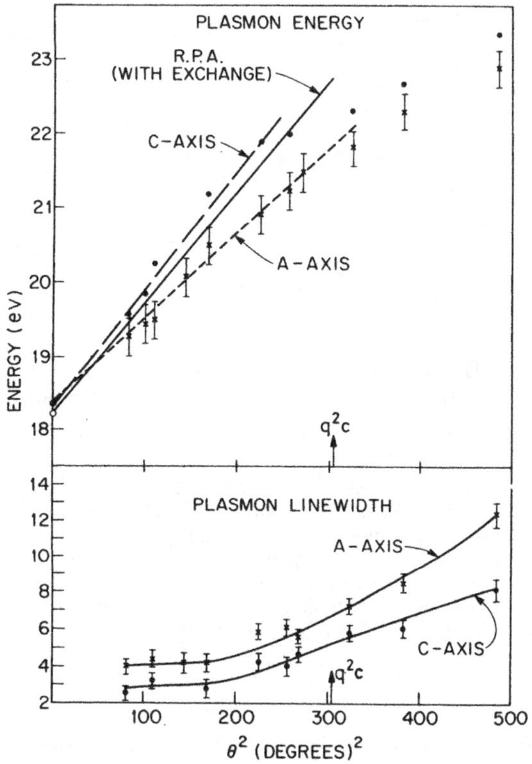

Fig. 6.2. Plasmon energy (above) and linewidth (below) of Be for $q \parallel c$ (dots) and $q \perp a$ (crosses) as a function of the square of the scattering angle θ measured with X-rays. The linewidth is the full width at half-maximum /6.8/. q see (3.14)

6.2 Influence of the Crystal Size on the Loss Width

Measurements of the loss width of the volume plasmons as a function of q up to $q \sim 0.3$ Å^{-1} ($\sim$ 2.5 mrad) demonstrated that in Al the linewidth decreases strongly with increasing q by about a factor of 3, see Fig. 6.3, whereas at larger q values the halfwidth increases as described by (6.1). This phenomenon occurs especially at fine crystalline material of a crystal size of about 60-70 Å /6.11,12/. If larger crystals are used as they are obtained by vaporisation without any precaution, only a small decrease of the width is observed at lower scattering angles (θ: 1-2 mrad) which changes into the positive dispersion, see Fig. 6.3. Similar behaviour has been found on Be, Si, and Ge /6.13/.

This effect can be better observed on Al, if the energy width of the primary electron is reduced, since the halfwidth of the Al loss has a rather small value of about 500 meV. Figure 6.4 demonstrates this size effect at the 15 eV loss of

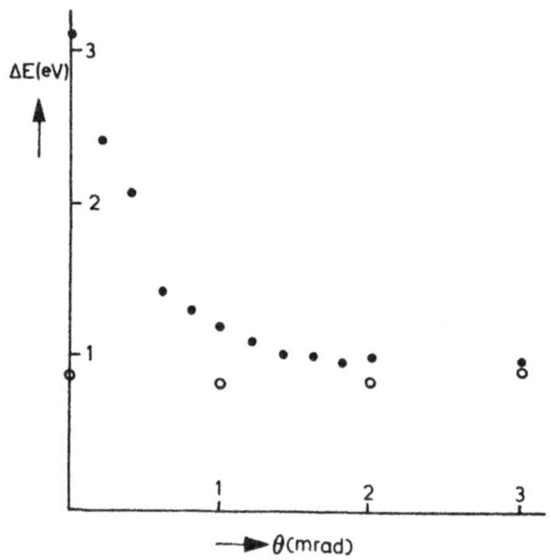

Fig. 6.3. The halfwidth of the 15 eV Al loss at small θ values. Primary electron energy 50 keV (•) fine crystalline material; (o) larger crystals, size about 200 Å /6.11/

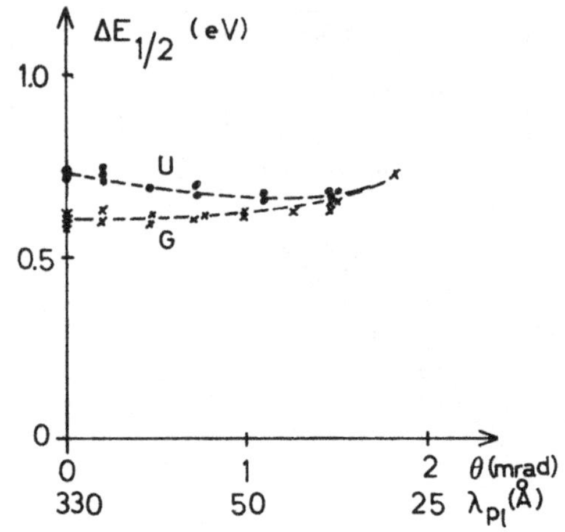

Fig. 6.4. Halfwidth of the Al 15 eV plasmon loss for annealed films (G) and for not annealed films (U) as function of the scattering angle θ. The corresponding plasmon wavelength λ_{Pl} is indicated. The primary electrons had an energy width $\Delta E^{o}_{1/2}$ of ≅ 100 meV /6.14/

Al at nonannealed (U) and at annealed (G) Al films observed with an energy width of 100 meV of the primary electron beam. The curve (U) still has a negative dispersion whereas the q dependence on a film heated for ∿ 2 hours at 400-500 °C (curve G) no longer has a negative slope. This demonstrates again that the negative slope at small q values is correlated in this case to the size of the crystals.

This observation can be explained in a plausible way by comparing the wavelength λ_{Pl} of the plasma oscillation with the mean size of the crystals D /6.11/: if one assumes that a plasma oscillation is additionally damped by crossing a grain boundary of two neighboured crystals, the damping increases the more, the more boundaries lie within the wavelength λ_{Pl} of a plasma oscillation or the more $\lambda_{Pl} > D$ is fulfilled. Since λ_{Pl} decreases with θ as

$$\lambda_{Pl} = 2\pi/q = 2\pi/k_{el}\cdot\theta \tag{6.2}$$

the damping should be reduced with larger θ values till λ_{Pl} gets comparable with D. Beyond these θ values the damping increases as observed at the other substances described by (6.1). Since a minimum of $\Delta E_{1/2}$ can be seen in Figs. 6.3 and 6.4 at $\theta \sim 2$ mrad, (6.2) gives with $k_{el} = 117$ Å^{-1} a value of D ∿ 25 Å which seems reasonable. A more detailed calculation has been given in /6.15/.

Measurements of the width of the *surface* peak of Al deposited on carbon films of different temperatures showed that the width of the surface peak has larger values (3-4 eV) at room temperature than at low temperatures of about -100 ^{o}C (1.5-2 eV). The width decreases with the scattering angle θ up to some 10^{-4} rad to a value of 1.5-2 eV; this behaviour is similar to that of the volume plasmons as we have just seen. Very probably the crystal size effect influences the damping of the surface losses too. It was not possible to decide whether the peak width increases again at larger angles θ as in the case of volume plasmons /6.22/.

6.3 Comparison with Theory

It is a fundamental question how to understand the short lifetime of the volume plasmons. For example in Al, which has one of the smallest loss widths of 500 meV, one finds using (1.14) that after about 10 oscillations their amplitude is reduced to 1/e which indicates a rather strong damping of the plasmons. The reason for this fact was unexplained for a long time, although it was proposed rather early that the excitation of interband transitions may be responsible for the short lifetime /6.16/.

Using the band structure of Si LOUIE et al. /6.17/ calculated successfully the dielectric function $\varepsilon_{GG'}(\omega,q)$ of Si and with it the loss function, its peak position, its halfwidth as well as its shape. Figure 6.5 compares these results with those found experimentally by energy loss experiments and by optical measurements. The agreement supports the assumption that the damping of the plasma waves for $q = 0$ is mainly determined by vertical interband transitions. This result represents an important progress in the physics of plasmons.

Already earlier calculations have been made for the alkali metals by PAASCH /6.20/. Table 6.2 shows the results of the position of the plasma loss, its halfwidth and its peak height, and the comparison with experimental values. There is in general good agreement except for K and Na. For Na the theoretical value of $\Delta E_{1/2}$ is about one-third of the experimental value; since the electron loss measurements are performed with an energy resolution of $\sim$ 0.1 eV /6.25/ the experimental value is certainly smaller than $\sim$ 0.3 eV. The experimental value of $\Delta E_{1/2}$ of K has been taken from optical work /6.26/.

Position and halfwidth of the volume plasmons of III/V compounds have been calculated recently /6.27/. The local field corrections improve the agreement of observed and calculated data similar as in Si /6.17/. This relatively good agreement does not exclude that other interactions, e.g., electron-electron correlations are relevant, a question which is not yet settled.

HASEGAWA /6.21/ attacked the problem of the loss width too; he took into account the effect of the periodic lattice potential and of phonons to calculate the q dependence of $\Delta E_{1/2}$ for small θ values and obtained

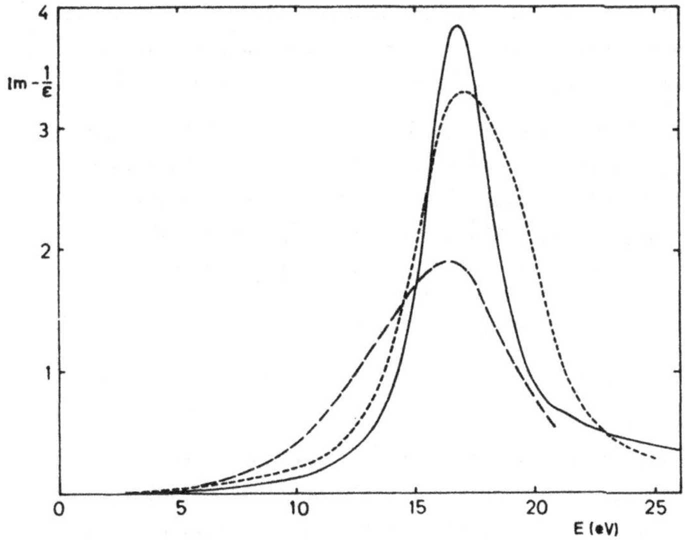

Fig. 6.5. The loss function of monocrystalline Si at 16.9 eV. Film thickness 1200 Å, E_0 = 50 keV. Solid line: experimental values /6.18/. Dashed line: calculated values /6.17/. Dotted line: calculated from optical data /6.19/

Table 6.2. Energy loss values $\Delta E_p(0)$, their halfwidth $\Delta E_{1/2}(0)$, and the maximum of the loss functions. Comparison of measured and calculated values

	ΔE_p [eV]				$\Delta E_{1/2}$ [eV]				$Im(-1/\varepsilon)_{max}$			
	exp.		theor.		exp.		theor.		exp.		theor.	
Li	7.08	/6.3/	7.30	/6.20/	2.2	/6.3/	2.55	/6.20/			3.9	/6.20/
Na	5.70	/6.3/	5.59	/6.20/	0.3	/6.25/	0.1	/6.20/				
							0.13	/6.21/	78	/6.20/	47	/6.20/
K	3.70	/6.3/	3.88	/6.20/	0.25	/6.26/	0.15	/6.20/	7.7	/6.20/	19	/6.20/
Rb	3.41	/6.23/	3.30	/6.20/	0.6	/6.23/	0.64	/6.20/	4.4	/6.20/	4.3	/6.20/
Cs	2.9	/6.23/	2.96	/6.20/	1.2	/6.23/	0.96	/6.20/	2.5	/6.20/	2.3	/6.20/
Al	15.0	/6.30/	15.5	/6.29/	0.53	/6.30/	0.44	/6.21/	24	/7.17/	54	/6.29/
							0.41	/6.24/			43	/6.24/
Mg					0.7	/6.3/	0.7	/6.24/				
					0.8	/6.28/						
Si	16.9	/6.18/	17.2	/6.17/	3.2	/6.18/	5.4	/6.17/	5.3	/6.18/	3.3	/6.17/
Ge	16.2	/6.18/	16.1	/6.27/	3.1	/6.18/	3.9	/6.27/	5.2	/6.18/	4.1	/6.27/

$$\Delta E_{1/2} = A + B\theta^2 + C\theta^4 + \ldots \tag{6.3}$$

in which A means the loss width at $\theta = 0$ and the factor B gives the q^2 dependence, see (6.1). Both coefficients A and B contain the contributions of the lattice potential and of the phonons, which is in general smaller. His calculations are restricted to very small q values and had the result that the value of B is negative in Al and Na. This has not yet been experimentally verified.

Recently the $\Delta E_{1/2}$ (q) dependence of the 15 eV Al loss has been calculated up to large q values /6.29,31/ and a good agreement has been found with the observed values, see Fig. 6.6, including the strong increase of $\Delta E_{1/2}$ at $q \sim q_c$, see Fig. 7.13, which indicates the decay of the plasmon into single particles. The theoretical dependence of $\Delta E_{1/2}$ in different crystal directions is not yet observed since the loss is too much enlarged at those q values at which the splitting of $\Delta E_{1/2}$ in the different crystal directions gets measurable.

Probably the fine structure in the $\Delta E_{1/2}$ (q) dependence calculated in /6.29/ is related to that observed in /6.30/.

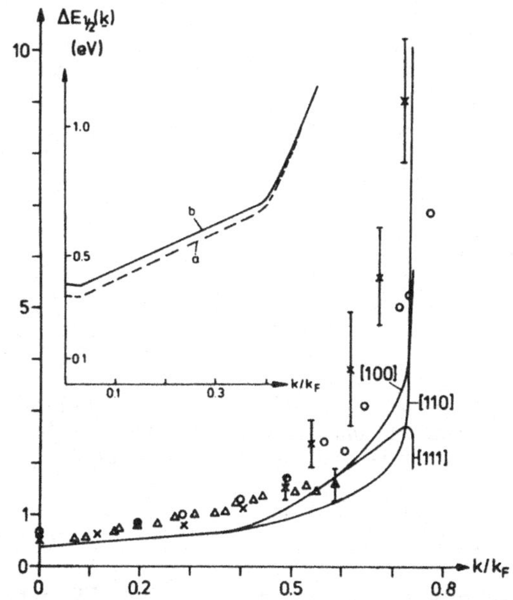

Fig. 6.6. The calculated wave vector dependence of the width of the volume plasmon peak of Al in different crystal directions (full line) /6.31/. Experiments on polycrystalline Al: crosses /6.32/, circles /6.25/, triangles /6.3/. The inset shows the influence of local field corrections in [100] direction, (a) without, (b) with correction

The anisotropy of $\Delta E_{1/2}(0)$ in the case of Be and Mg has been calculated too, /6.24/; the halfwidths of the Mg loss parallel to the a axis and parallel to the c axis are 0.66 and 0.86 eV, respectively, whereas the observed value is equal for both directions at $q \cong 0$ and amounts to 0.7 ± 0.1 eV; the latter is the mean value of the calculated figures. The theoretical values of Be parallel to a and parallel to c are 2.05 and 2.29 eV, respectively, whereas observed ones are not reported at $q = 0$. Extrapolation to $q = 0$ in Fig. 6.2 gives a value between 2 and 4 eV in approximate agreement. For $q > 0$ however the observed $\Delta E_{1/2}$ $(q \parallel a)$ is larger than $\Delta E_{1/2}$ $(q \parallel c)$, see Fig. 6.2.

Besides the absorption of the plasmon energy by an interband transition which produces most of the linewidth of the volume loss as we have seen there exists the possibility that the plasmon picks up a reciprocal lattice vector G and reaches thus the continuum band of the single particle excitation where it decays into an electron-hole pair /6.21/, Fig. 6.7. In general the interband contribution is larger than the contribution of this process.

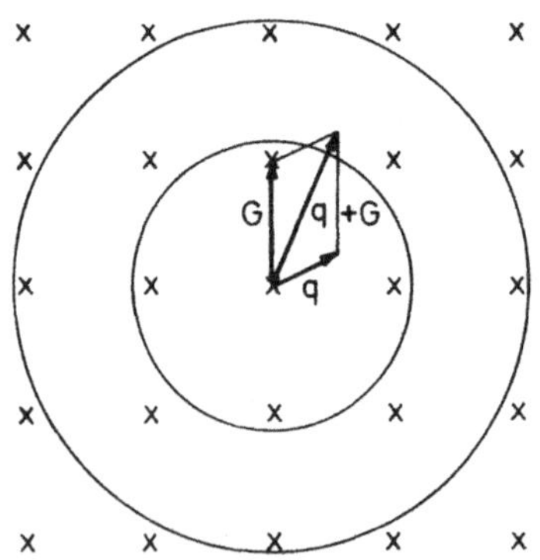

Fig. 6.7. A plasmon of momentum $\hbar q$ picking up $\hbar G$ (G reciprocal lattice vector) reaches the region of the single particle excitation and decays. The inner circle has the radius of q_c whereas the outer $q_c + 2q_F$, see also the one-dimension Fig. 1.2

Instead of picking up the quantity $\hbar G$ the plasmon can exchange its momentum with a phonon; the probability of this process however is in general negligible.

The general rather good agreement between observation and calculation means that the main part of the q dependence of $\Delta E_{1/2}$ is accounted for by nonvertical electronic transitions fitting into the band structure and that the number of such transitions increases with q so that damping becomes larger. It may happen that the number of transitions decreases with increasing q at first and increases at higher q values. This seems to be the case in Li /6.3/ where a minimum in the $\Delta E_{1/2}$ (q) dependence results from calculation /6.31/.

It is probable that this dependence will show a fine structure if detailed measurements of $\Delta E_{1/2}$ (q) are made, since the number of nonvertical transitions per energy interval will show a certain variation; perhaps the results in /6.30/ indicate such an effect.

Before it has become evident that the plasmon halfwidth is due to the interband transitions in the crystal an interpretation has been proposed within the concept of the free electron gas. Here the only process leading to a decay of the collective oscillation consists in the transfer of the energy $\hbar\omega_p$ and the momentum $\hbar q$ to one electron-hole pair, a process which defines the Landau edge and occurs only for $q \geq q_c$, q_c the cutoff vector, see Sect. 1.4. Below this limit damping is only possible in the concept of the free electron gas by a second-order process; this is the creation of two electron-hole pairs or multipairs to fulfill momentum and energy conservation /6.33/. Earlier work ascribed the interpretation of the observed damping to this second-order process /6.34/ but the quantitative treatment was not correct; the second-order production of pairs is much too small to explain the observed values /6.35,36/.

7. The Wave Vector Dependency of the Energy of the Volume Plasmon

7.1 Dispersion at Small Wave Vectors ($q < q_c$)

As we have seen the peak of the loss function $\mathrm{Im}(-1/\varepsilon)$ lies at the plasma frequency ω_p for $q \to 0$ and for small ε_2. A q dependent peak position is obtained theoretically by introducing a q dependent ε into the loss function. In the case of a free electron gas this has been discussed in Chap. 1. For $q \ll q_F$ we have obtained (1.17). With $q = k_{el}\theta$ we can write (1.19) as

$$\Delta E(\theta) = \Delta E(o) + 2E_o\alpha\theta^2 \tag{7.1}$$

or

$$\Delta E(q) = \Delta E(o) + \frac{\hbar^2}{m}\alpha q^2 \tag{7.2}$$

with

$$E_o = \frac{\hbar^2}{2m}k_{el}^2 \quad \text{and} \quad \alpha = \frac{3}{5}\frac{E_F}{\Delta E_p} \ . \tag{7.3}$$

In order to observe the q dependence, the value of the energy loss $\Delta E(\theta)$ as a function of θ is measured in an arrangement shown in Fig. 3.3. The experimental dependence can now be compared with the calculated dispersion relation.

Polycrystalline substances. The first experiments to measure the q dependence in transmission of thin films were made by WATANABE /7.1/, see also /7.2/. An example of recent measurements on polycrystalline materials shows, Fig. 7.1 /7.3/, the loss position comes out in a first approximation as linear function of q^2

$$\Delta E(q) = \Delta E(o) + A q^2 \tag{7.4}$$

nearly up to q_c; for some q_c values see Table 7.1. Comparison of the experimental slope

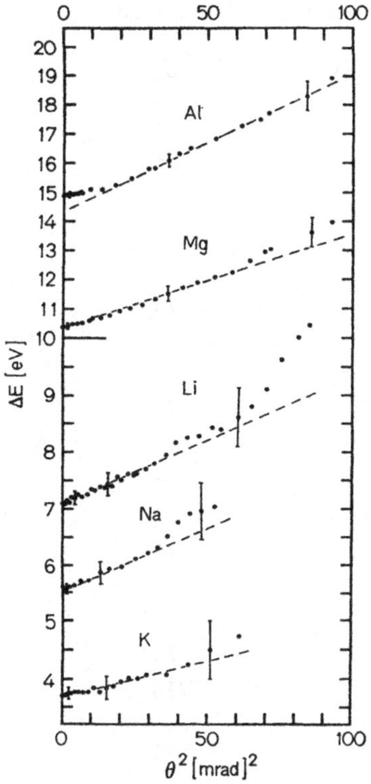

Fig. 7.1. Dispersion of the volume plasma oscillations in different polycrystalline materials. The scattering angle θ and wave vector q are related by $q = 117\ \text{Å}^{-1} \cdot \theta$ for 50 keV electrons /7.3/

Table 7.1. Some values of the cutoff vector q_c and θ_c and the Fermi vector q_F for the substances of Fig. 7.1 $q_c = 117\ \theta_c\ \text{Å}^{-1}$ at 50 keV electrons

	$q_c [\text{Å}^{-1}]$	θ_c [mrad]	$q_F [\text{Å}^{-1}]$
Al	1.3	11.1	1.75
Mg	1.0	9.0	1.37
Li	0.9	8.0	1.11
Na	0.8	7.4	0.92
K	0.8	6.7	0.75

Table 7.2. Values of the dispersion coefficient α, observed and calculated

Substances	α(exp.)		α_L(calc.) (Eq. 1.20)	Ref.
Li	0.26 ± 0.04		0.35	/1/
	0.22 ± 0.03			/2/
Na	0.22 ± 0.04		0.32	/2/
	0.29 ± 0.02			/1/
	0.25			/3/
K	0.12 ± 0.04		0.29	/2/
	0.15 ± 0.05			/1/
Be	0.44		0.47	/4/
	0.42			/5/
	0.41 X-rays			/6/
Mg	0.39 ± 0.01		0.39	/7/
	0.32 ± 0.03			/2/
	0.35 ± 0.02			/18/
Ca	< 0.03			/1/
	0.37 ± 0.03			/1/
	0.03			/8/
	0.38 ± 0.04			/8/
Sr	< 0.03			/1/
	0.12 ± 0.04			/1/
	0.04			/8/
	0.25 ± 0.1			/8/
Ba	0.24 ± 0.05			/1/
	0.25 ± 0.04			/8/
Ga	0.45 ± 0.05	cryst., liq.	0.43	/9/
Sn	0.54 ± 0.05	cryst., liq.	0.42	/9/
Sb	0.48 ± 0.03	polycryst.	0.44	/12/
	0.51 ± 0.05	polycryst.		/10/
	0.37 ± 0.03	oriented		/4/
	0.39 ± 0.03	oriented		/12/
	0.44 ± 0.07	oriented		/10/
	0.43 ± 0.08	am.	0.42	/10/

Table 7.2. (continued)

Substances	α(exp.)	α_L(calc.)(Eq. 1.20)	Ref.
GaP	0.61	0.45	/11/
InAs	0.48	0.43	/11/
InSb (see also Table 7.4)	0.38	0.41	/11/
GaSb	0.28	0.43	/11/
GaAs	0.44	0.44	/11/
Si (see also Table 7.4)			
Ge	0.38	0.44	/5/
Graphite (see Fig. 7.2)			
Ag (see Fig. 7.5)			
Al (see Table 7.3 and 7.4)			
In (see Table 7.3)			
PbS, PbSe, PbTe (see Table 7.4)			

References to Table 7.2

/1/ T. Kloos: Z. Phys. 265, 225 (1973)
/2/ C. Kunz: Z. Phys. 196, 311 (1966)
/3/ O. Sueoka: J. Phys. Soc. Jpn. 20, 2203 (1965)
/4/ J. Geiger: Z. Naturforsch. 17a, 696 (1962)
/5/ H. Watanabe: J. Phys. Soc. Jpn. 11, 112 (1956)
/6/ G. Priftis: Phys. Rev. B 2, 54 (1970)
/7/ P. Schmüser: Z. Phys. 180, 105 (1964)
/8/ J. Langkowsky: J. Phys. D, Appl. Phys. 8, 2058 (1975)
/9/ K. Zeppenfeld: Z. Phys. 223, 32 (1969)
/10/ B. Bartning: Opt. Commun. 4, 404 (1972)
/11/ C. v. Festenberg: Z. Phys. 214, 464 (1968)
/12/ I. Cazaux, R. Vilanove: C.R. Acad. Soc. Paris B 263, 460 (1966)
/13/ J. Stiebling, H. Raether: Phys. Rev. Lett. 40, 1293 (1978)
/14/ M. Urner-Wille, H. Raether: Phys. Lett. 58A, 265 (1976)
/15/ H.I. Höhberger, A. Otto, E. Petri: Solid State Commun. 16, 175 (1975)
/16/ U. Büchner: Phys. Status Solidi (b) 83, 493 (1977)
/17/ R. Manzke: J. Phys. C, Solid State 11, L349 (1978)
/18/ C.H. Chen: J. Phys. C, Solid State Phys. 9, L321 (1976)

Remarks to Table 7.2

In the spectra of Ca and Sr two volume plasmons are observed; their α values are given in brackets.

The α coefficients of the Ga and Sn plasmons are the same in the crystalline and in the liquid phase.

Electron-electron corrections have not been made, since different relations can be chosen, see Sect. 7.2 and Table 7.5, with rather different results.

$$A = \frac{\hbar^2}{m} \alpha_{exp} = 7.6\ \alpha_{exp}\ [\mathrm{eV\ Å^2}] \tag{7.4a}$$

is often made with the calculated α_{calc}, see (7.4), developed for small q values and a free electron gas. The α data show rather different agreement, see Table 7.2.

As mentioned in Chap. 3.2 and Fig. 3.5 the dispersion of the volume plasmon can be observed with X-rays, as in Be and in Li /7.4-6/; the α values of Li of 0.34 - 0.36 obtained with X-rays do not agree well with ELS data.

Looking more exactly at the 15 eV loss in polycrystalline Al at small q values KLOOS /7.3/ found that the curve $\Delta E(\theta^2)$ starts with a linear part which turns over to a steeper straight line (in the limits of error), see Fig. 7.9, so that two values of α can be derived; see Table 7.3. This type of dependence at Al has been confirmed by several authors /7.7-9/. A similar bend-off in the dependence $\Delta E(q)$ has been found at In /7.9/ and can be supposed in Mg /7.9/.

Table 7.3. Observed dispersion coefficient α of Al and In in different q regions

	α	q	Ref.
Al	0.13 - 0.17	< 0.4 Å^{-1}	/7.3/
	0.21	< 0.5 Å^{-1}	/7.9/
	0.45	> 0.5 Å^{-1}	/7.3/
	0.46	> 0.5 Å^{-1}	/7.9/
In	0.40	< 0.5 Å^{-1}	/7.9/
	0.66	> 0.5 Å^{-1}	/7.9/

As is well known the energy loss of the volume plasmon splits at larger q values into an undispersed loss $\hbar\omega_p(0)$ and the dispersed one. The first loss is the result of a two-step process: 1) elastic scattering into the angle θ by structural imperfections of the film, by coating with amorphous oxide or by phonons /4.21/; 2) inelastic scattering without considerable momentum transfer. This phenomenon takes place in Al at q values between 0.9 and 1.1 Å^{-1} and leads to an increased error in the determination of the dispersion loss position in this small q region, see, e.g., /7.9/. One can assume that phonons are not of importance for process (1) since a change of temperature from room to helium temperature does not influence the intensity relation of the inelastic to the elastic peak, /7.17/.

The dispersion relation has been proved not only with electrons and X-rays, but also with light. As mentioned in Sect. 3.3, light of frequency ω_p and higher of non-normal incidence reflected at a smooth metal surface can excite volume plasmons. Resonances (ω_{res}) (maximum excitation of the plasma oscillation) appear when an odd number n of half wavelengths λ of the volume plasma oscillation fits into the film thickness d, i.e.,

$$n\lambda = 2d$$

or

$$q = \frac{2\pi}{\lambda} = \frac{\pi n}{d} \tag{7.5}$$

Reflection of light of varying frequency at thin potassium films verified that /7.10/

$$\omega_{res} \sim n. \tag{7.5a}$$

Since the thickness of the very thin potassium film is rather difficult to determine and thus the values of q (7.5), the value of the coefficient α could not be checked. The thickness or the wave vectors q can be calculated by assuming that $\alpha = \frac{3}{5} E_F/\Delta E_p$ (7.3) is valid, but as has been shown in /7.3/ the observed value of α is nearly two times smaller than the calculated one, so that this assumption is not allowed.

Monocrystalline substances. Observations on monocrystalline solids showed that the α values are dependent on the direction of q in the crystal.

The first observation of an *anisotropy* in the ELS, i.e., a dependence of the structure of the loss spectrum of the direction of $\vec{q}$ (at $|\vec{q}| \sim 0$) inside the crystal at a fixed $|q|$, has been reported on graphite /7.11/. ZEPPENFELD could explain the changes of the structure of the spectrum with the variation of the angle between the primary electron beam and the c axis of the graphite crystal see (2.15a).

Further detailed experiments revealed in addition that there exist a dependence of ΔE on the *value* of $|q|$, different in different directions of $\vec{q}$. Such an anisotropy of α has been found in the different directions $\overrightarrow{\Gamma P}$ and $\overrightarrow{\Gamma Q}$ in the plane perpendicular to the c axis of graphite, see Fig. 7.2, curve (a) /7.12/. A difference of α in the direction of the c axis and perpendicular to it exists too but this is not surprising since the correlated loss positions are different due to the anisotropy, see Fig. 7.2, curve (b), see also /7.13/.

In further experiments with electrons and X-rays this anisotropy has been observed in other substances too, see Fig. 7.3. Experiments with X-rays on the hexagonal beryllium showed different q dependencies in and perpendicular to the c axis /7.14/, Fig. 6.2. Electron loss experiments with epitaxially grown and monocrystalline thin films of Al, Si, InSb, and lead chalcogenids demonstrated this dependence of the dispersion of the volume loss on the direction of q in the crystal. The change of α with q in different directions is more pronounced in semiconductors

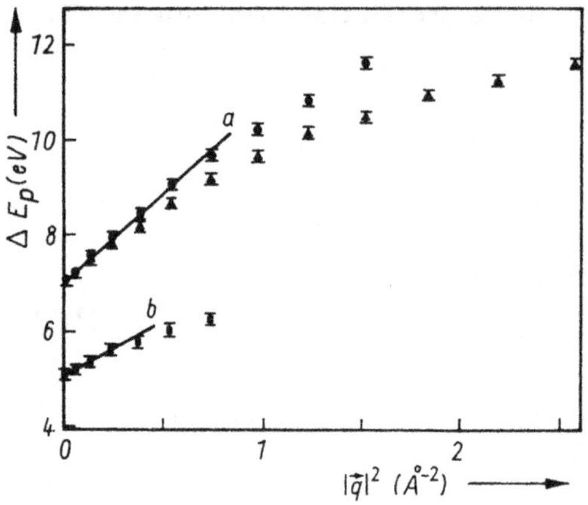

Fig. 7.2. Dispersion of the energy losses of low ΔE in graphite in different crystal directions /7.13/. The dispersion coefficient $\alpha(\overrightarrow{\Gamma Q})$ (•) of the π plasmon (7 eV) [upper curve (a)] has a value of 0.43 for $q < \sim 0.7$ Å^{-1}, whereas $\alpha(\overrightarrow{\Gamma P})$ (▲) is smaller than $\alpha(\overrightarrow{\Gamma Q})$. $\alpha(\overrightarrow{\Gamma A})$ of the 5 eV loss has a value of 0.28 [lower curve (b)]. The crystal directions are explained in Fig. 7.20

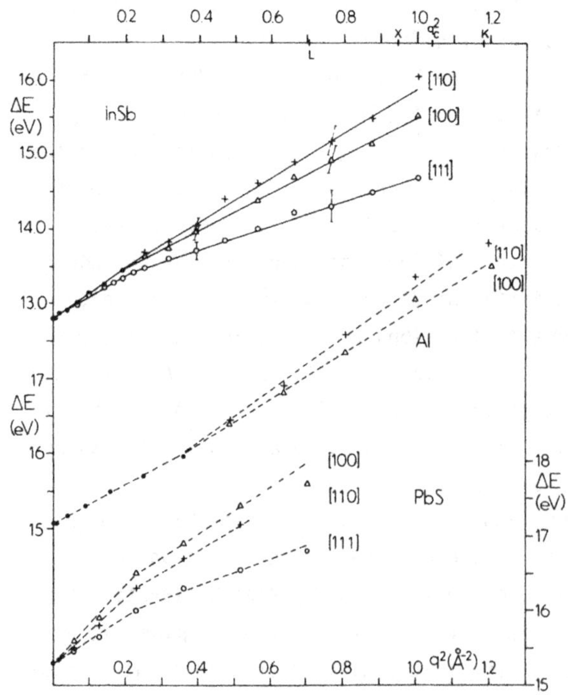

Fig. 7.3. Plots of $\Delta E_p(q^2)$ of different monocrystalline solids in different crystal directions /7.15/

than in metals, see Fig. 7.3. Whereas in the metals Al and In the α values are higher at larger q than those at smaller q values, the reverse is observed at semiconductors. The values of the anisotropic dispersion coefficients are collected in Table 7.4. The two figures mean α_{exp} for $q < \bar{\bar{q}}$ and $q > \bar{\bar{q}}$.

Table 7.4. Anisotropic dispersion coefficients of different substances

	$\alpha_{exp.}$			$\alpha_{calc.}$	Ref.
	$\vec{q} \parallel (100)$	$\vec{q} \parallel (110)$	$\vec{q} \parallel (111)$		
Al $\bar{\bar{q}}^2$: 0.3 - 0.4	0.2; 0.40	0.2; 0.46	0.2; 0.43	0.44	/7.23/
InSb $\bar{\bar{q}}^2$: 0.15 - 0.25	0.44; 0.33	0.44; 0.39	0.36; 0.21	0.41	/7.15/
PbS $\bar{\bar{q}}^2$: 0.2 - 0.3	0.62; 0.41	0.53; 0.38	0.39; 0.24	0.40	/7.24/
PbSe $\bar{\bar{q}}^2$: 0.1 - 0.2	0.67; 0.42	0.54; 0.36	0.50; 0.20	0.39	/7.24/
PbTe $\bar{\bar{q}}^2$: 0.15 - 0.25	0.66; 0.37	0.52; 0.36	0.46; 0.30	0.38	/7.24/
Si $\bar{\bar{q}}^2$: 0.25 - 0.35	0.41; 0.41	0.41; 0.41	0.41; 0.32	0.45	/7.39/

$\bar{\bar{q}}^2$ transition region, in $Å^{-2}$, in which α changes its value.

The measurements of the dispersion relation in monocrystals can meet difficulties, if one observes in directions [hkl] in which the diffraction maximum (hkl) has appreciable intensity. The inelastic electrons of the reflex as well as of the primary beam contribute then to the loss spectrum in a given θ. In the case of Si the intensity of the diffraction peak (111) may be a factor of 10 smaller than the primary beam. The consequence is shown in Fig. 7.4: here the dispersion of the Si 16.9 eV loss is plotted along the direction [111]; the reflex (111) lies at 2.00 $Å^{-1}$. At $q \geq 1.4$ $Å^{-1}$ the dispersed loss from the forward direction has practically disappeared, whereas the loss coming from the reflex can no longer be detected below q = 1.2 $Å^{-1}$. At 1.27 $Å^{-1}$ one can discern the two losses from both beams, the primary and the reflected. In the <100> direction of Si the dispersion

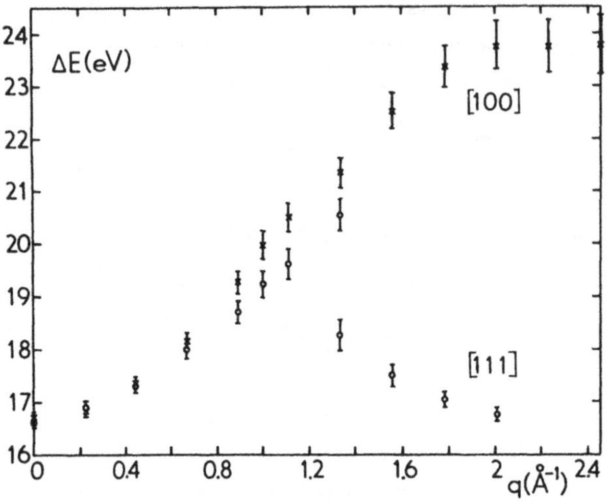

Fig. 7.4. Dispersion of the 16.9 eV loss in the [111] and [100] direction. In the [100] direction the loss can be observed up to $q \sim 2.4$ Å^{-1}, whereas in the [111] direction the loss produced by the (111) diffraction maximum dominates for $q > 1.4$ Å^{-1} /7.16/

can be measured up to 2.2 Å^{-1} since the reflex (200) lies at 4.63 Å^{-1}. This circumstance allows one to observe the $\Delta E(q)$ dependence up to larger values of q, see /7.39/, in the [100] direction.

Plasma oscillations of bound electrons. If the electrons can no longer be regarded as free and the gap energy is comparable with the collective oscillation energy, the situation is less simple than, e.g., in Ge, Si, III-V compounds, etc.. A typical case is Ag; here in the low-energy region (3-4 eV) the surface loss ($\varepsilon = -1$, $\hbar\omega_s = 3.6$ eV) and the volume loss ($\varepsilon = 0$, $\hbar\omega_p \doteq 3.78$ eV) are close together due to the fact that the interband transition from the 4d band to the Fermi level (≥ 3.9 eV) is not small compared to 9 eV, the plasma energy of the free electron gas with a density of one valence electron per atom of silver. Interband transitions of higher energies /7.21a/ displace also the theoretical value to lower ones. The resulting plasma frequency comes out as 3.78 eV. The hybrid character of this oscillation is demonstrated by the small dispersion of the 3.78 eV loss. The plot of ΔE_p as function of q^2 for the 3.78 eV gives a straight line with a value of

$$\alpha = \frac{3}{5}\frac{E_F}{\hbar\omega_p} = 0.37$$

($E_F = 5.5$ eV and $\hbar\omega_p = 9$ eV) for the free electron gas whereas the experimental value comes out as $\alpha \cong 0.7$, see Fig. 7.5. The effect of the lattice is thus evi-

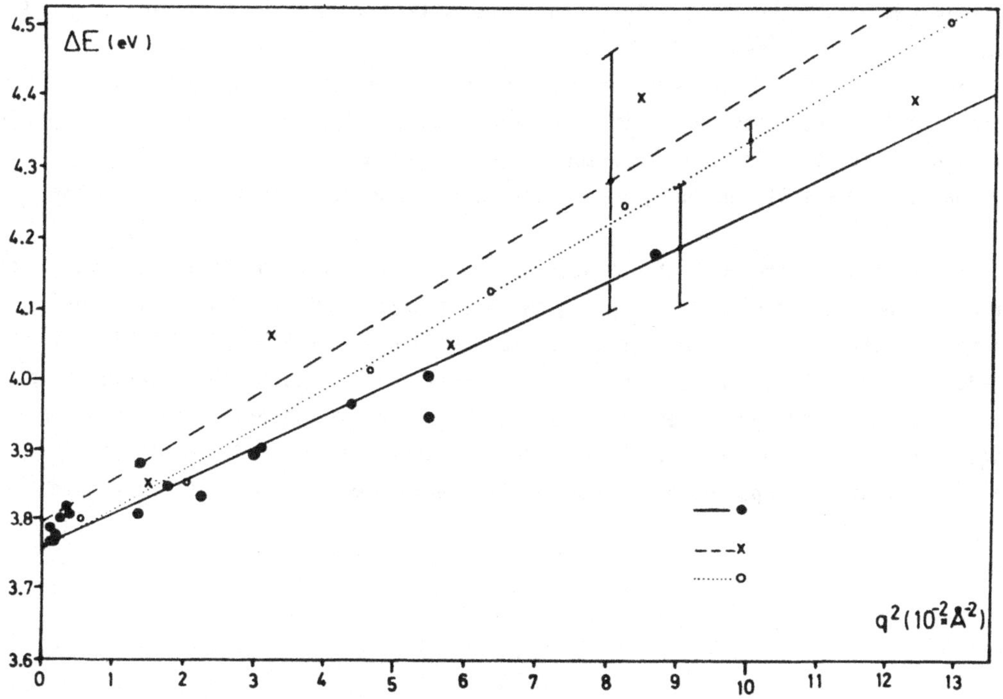

Fig. 7.5. Dispersion $\Delta E(q^2)$ of the 3.76 ± 0.03 eV energy loss in polycrystalline silver measured by different authors /7.18/.
(●)α = 0.63 ± 0.09 /7.18/; (o)α = 0.76 ± 0.03 /7.20/;
(x)α = 0.80 ± 0.16 /7.19/; from optical data /7.21/.
α = 0.78 ± 0.01 (not included in the figure)

dent. It is interesting to notice that the intensity of the loss decreases faster than q^2, an effect which is very probably due to rapid damping produced by ΔE_p getting equal to the energy of an interband transition. A similar situation was assumed to explain the increase of the width of the loss peak with increasing the Ga concentration in a Ag-Ga alloy observed at small q values /7.21b/.

In the complex spectra, e.g., of the alkali halides the dispersion of different losses has been measured. Of special interest are the peaks interpreted as plasmon losses, e.g., 25.3 eV LiF; it shows no increase of energy as function of the wave vector q up to q ∿ 10 mrad (50 keV) electrons. The dispersion coefficient A, see (7.4), is at least ten times smaller than that of Al /7.22/. This indicates that we have not to do with quasi free electrons in the sense that the calculated plasma energy $\hbar\omega_p$ is large compared with the binding energy of the electrons, see also Chap. 5, c.

Measurements at the relatively sharp excitonlike losses 13.5 eV LiF and 10.3 NaF had similar results, see Sect. 7.3.

The 62.0 eV LiF loss showed up to 15 mrad (50 keV electrons) no change of its energy /7.54/. The behaviour of this nearly single particle excitation is determined by the form of the band structure, see Sect. 7.3.

For further results on the q dependence in the case of strongly bound electrons see Sect. 7.3.

Comparison with theory. A first step to derive the dispersion of the plasma frequency has been done by calculating the dielectric function of Si taking into account the effect of band structure, see WALTHER and COHEN /7.25/. The local field correction has been neglected ($\varepsilon_{GG'}$ is replaced by ε_{00}) so that quantitative agreement could not be achieved; the progress however is evident if one compares Figs. 7.6 and 7.7. In Figure 7.6 the Lindhard loss function $\mathrm{Im}\{-1/\varepsilon(\omega,q)\}$ is displayed: the large humps represent the single electron excitation peaks discussed

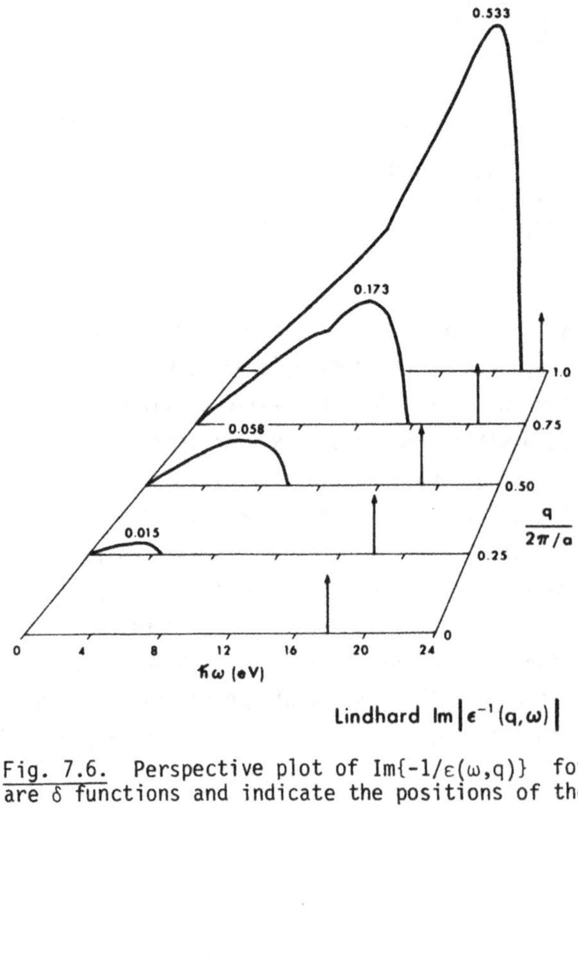

Fig. 7.6. Perspective plot of $\mathrm{Im}\{-1/\varepsilon(\omega,q)\}$ for a free electron gas. The arrows are δ functions and indicate the positions of the volume plasmons /7.25/. a = 5.4 Å

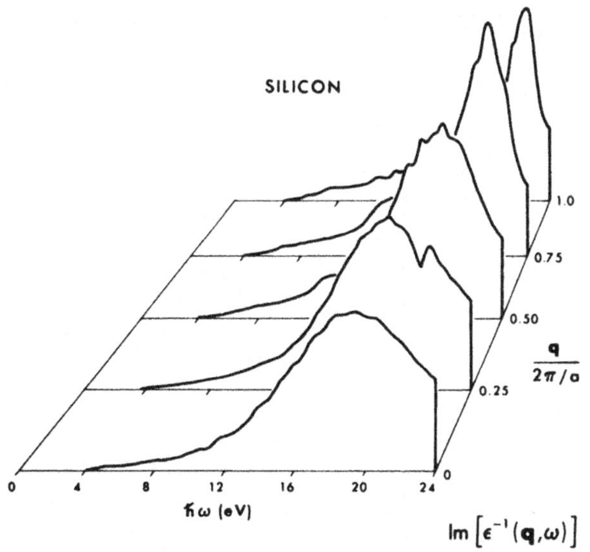

Fig. 7.7. Perspective plot of Im{-1/ε(ω,q)} for silicon /7.25/

in connection with Fig. 7.14; the arrows indicate the δ functions of the collective excitation. Figure 7.7 shows the result of the calculation which considers the influence of the interband transition. Besides the finite height of the Im(-1/ε) peak as well as the finite width instead of δ functions, already discussed in Sect. 6.3, the plasmon energy shows a characteristic q dependence, different from that of a free electron gas, see Fig. 7.6. More detailed calculations of the q dependence of the Si loss have not been accomplished up to now.

For potassium and sodium the dispersion relation ΔE(q) has been calculated in different crystal directions by HAGUE and KLIEWER /7.25a/ taking into account the band structure. A strong anisotropy is obtained in the [100] and [110] directions. Measurements on monocrystalline alkali metals however do not yet exist to compare the results with calculations.

Recently calculations have been published on the dispersion of the 15 eV Al loss by BROSS /7.26/. Applying the theoretical concept discussed in Chap. 2 (RPA dielectric function, one electron energy, wave functions derived by a pseudopotential approximation) the anisotropy of the dispersion relation of Al has been calculated. The results are displayed in Fig. 7.8. It compares the observed values (full circles) obtained as the difference of the energy positions of the losses in the [110] direction and in the [100] direction with respect to those of the [100] direction. The theoretical values are those calculated with the ε_{00} dielectric function without local field correction (dashed line) and with exchange correction (full line).

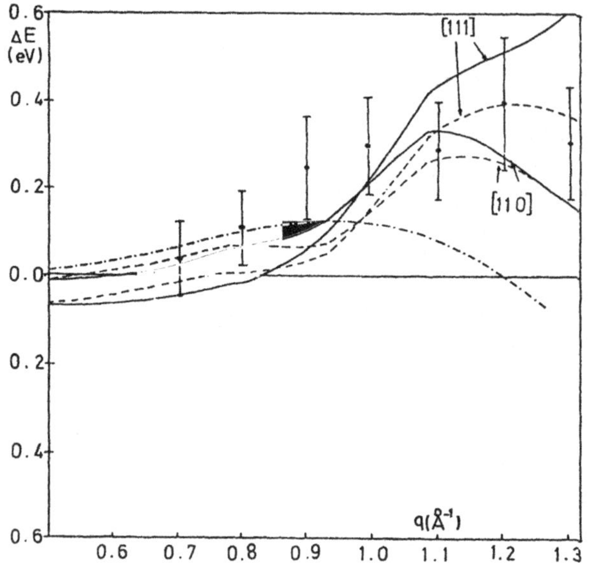

Fig. 7.8. Anisotropy of the dispersion of the 15 eV volume plasmon /7.26/. The difference of the plasmon positions between the [110] and [111] directions and the [100] directions as function of q. Solid line: with Hubbard correction. Dashed line: RPA. Points experimental values /7.23/

At $q < 0.7$ Å^{-1} a separation of the losses in different directions is not possible, the difference being too small; at larger q values it increases quickly and reaches values of $\sim$ 0.5 eV. The losses of the [110] and [111] directions could not be separated experimentally. Observed and calculated values are in good agreement. It is thus shown that the anisotropy $\Delta E(q)$ can be explained within the RPA including band structure using one electron energies and wave functions determined by a pseudopotential approximation.

These values calculated for monocrystalline Al can now be used to calculate the dispersion of polycrystalline Al which had been measured frequently. Averaging the ΔE values of Al of the principal crystal directions one obtains the dependence in Fig. 7.9 given by the dotted curve in the Lindhard approximation, by the dashed dotted curve using the Cohen-Ehrenreich function (interband transition considered), and by the full line with the Hubbard exchange correction. These curves describe the experimental curve rather well although the absolute values differ about a half up to $\sim$ 2 eV depending on q.

The important result of these theoretical considerations is that one is not far away from agreement between observation and calculation which means that the basic physical explanation is right. The bending off, i.e., the transition of a nearly linear part at small q values into another roughly straight line of steeper slope,

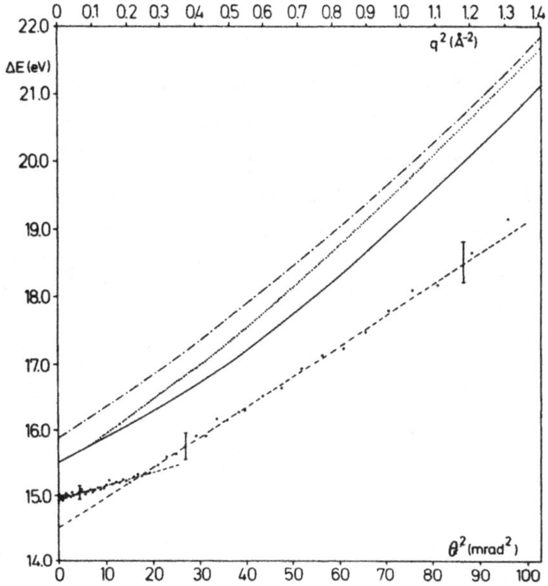

Fig. 7.9. The dispersion curve of the 15 eV plasmon of polycrystalline Al as function of q^2 /7.26/. Dash-dotted: Lindhard function. Dotted: Cohen-Ehrenreich function. Full: the same with HUBBARD correction /7.27/. Points: experimental values /7.9/

can be recognized in the calculated curves. The slopes themselves show still a certain discrepancy.

Other approximations of the electron-electron correlation have been proposed, but they did not remove the discrepancies, see Table 7.5. Recent calculations /7.46/ have given for Al the value $\alpha(q \to 0) = 0.42$, whereas the observed value is $\alpha_{exp} = 0.21$.

Conclusion. Resuming we see that the *energy* of the volume plasmons is given in a first good approximation by $\hbar\omega_p$ (1.8) basing on the free electron gas concept; the band structure corrects these values slightly, see Tables 5.1 and 6.2. Similarly their q dependence is approximately derived from this concept, see (1.18); the band structure correction displaces the $\omega(q)$ curve in the right sense, see Al (Fig. 7.9). The role of the electron-electron correlation is open. In contrast the plasmon *width* and its q dependence is basically determined by the band structure and not included in the free electron concept.

Table 7.5. Dispersion coefficient with different exchange corrections /7.29a/

	$\alpha_{calc.}$				$\alpha_{exp.}$	
	Lindhard	/7.27/	/7.28/	/7.29/		
Al	0.44	0.39	0.36	0.27	0.2	$q \sim < 0.5$ Å^{-1}
					0.45 ± 0.04	$q \sim > 0.5$ Å^{-1}
Mg	0.39	0.33	0.30	0.20	0.32 ± 0.03	

7.2 Dispersion at Large q Values

Experimental results. It is an interesting problem to study the behaviour of the plasma loss at values of q higher than q_c. The problem is whether the loss persists beyond q_c, what is its nature, and how its dispersion looks.

The first experiments were done with X-rays by PRIFTIS et al. /7.31/, see Chap. 3. Figure 7.10 reproduces the dependence of the position of the inelastic maximum on the scattering angle θ in Be, starting at ΔE_p = 19 eV and increasing

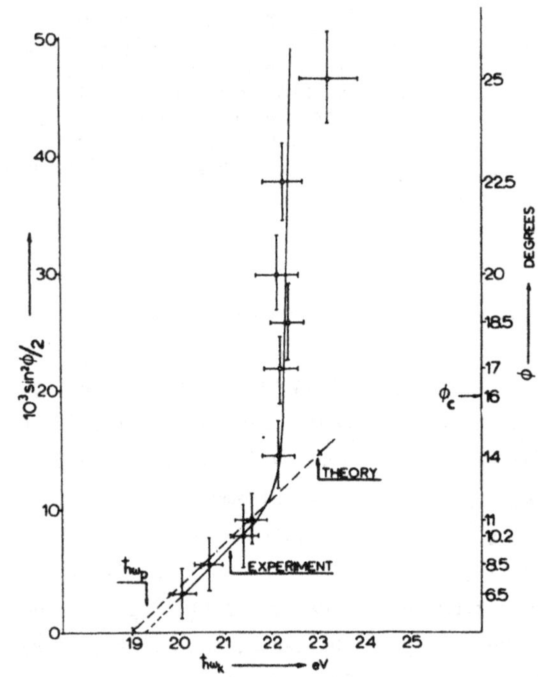

Fig. 7.10. Dispersion curve of the bulk plasmon of Be, measured with X-rays, λ = 1.392 Å (Cu K_β) /7.32/

linearly with θ^2 /7.32/. In Fig. 7.10 θ is called ϕ. At about 22 eV the loss position changes practically no more up to angles of about 25^0 ($\theta_c = 16^0$). The break in the linear dependence on θ^2 takes place between 11^0 and 14^0 still before θ_c has been reached. The general shape of the dispersion of the Be plasmon has been verified in further experiments /7.33/. Experiments on Li (ΔE_p = 7.9 eV) had a similar result; a different behaviour came out insofar as the inelastic peak retains its linear dependence on θ^2 up to 1.0 Å^{-1}; q_c = 0.9 Å^{-1} /7.34,35/.

$\Delta E_p(q)$ for $q > q_c$ has been studied with fast electrons by ZACHARIAS on Al /7.36/ too. The results on Al demonstrated a rather similar behaviour as obtained on Be: the break of the linear function of $\Delta E_p(\theta^2)$ into a curve of smaller slope at $q > q_c = 1.3$ Å^{-1}, see Fig. 7.12, crosses (+). A similar behaviour of the dispersion curve has been measured on Al in /7.37/, in agreement with the results of /7.36/.

Similar experiments were published on Al which had a rather different result insofar as the strong break in the slope of the $\omega(q^2)$ dependence is not observed /7.38/. see Fig. 7.12. These data are obtained by another evaluation of the loss spectrum in respect to the multiple scattering corrections. Experiments in ultrahigh vacuum may contribute to clear up this difference of the results since the lack of an oxide on the Al film will reduce the multiple scattering influence.

Further experiments were performed on Si monocrystals /7.39/ which are in general very weakly oxidized, as can be recognized by the weak oxide loss at about 6 eV. Here multiple scattering effects are less important. Figure 7.11 demonstrates

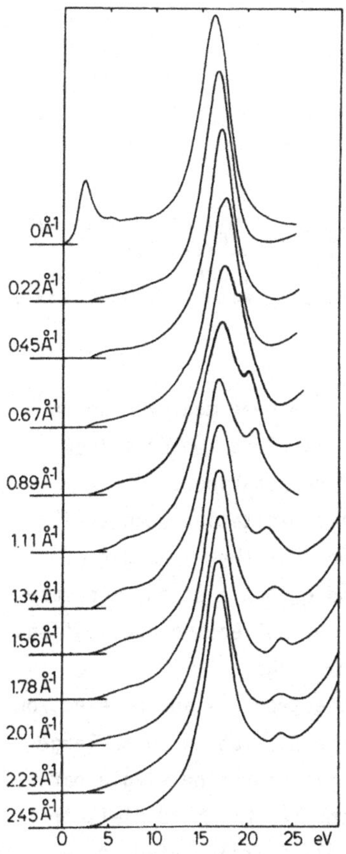

Fig. 7.11. Intensity of the loss energy for Si (arbitrary units) as a function of the wave vector up to $q = 2.5$ Å^{-1} ($q_c = 1.2$ Å^{-1}). The direction of q lies parallel to [100]. The peaks of the volume loss are normalized to the same height /7.39/. The large peak at $q > 1$ Å^{-1} is due to process (1), see page 75

how the 16.9 volume loss disperses up to about q_c: the slope of the $\omega(q^2)$ becomes smaller and for $q > q_c$ the position of the loss peak remains nearly constant at about 24 eV, in agreement with the dependence observed in Al /7.36/.

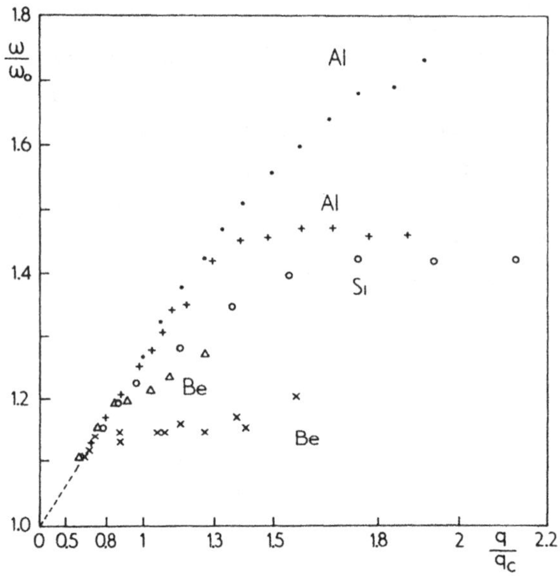

Fig. 7.12. Normalized experimental dispersion curve ω/ω_o ($\omega_o = \omega_p$) of several substances up to large q vectors /7.39/. Solid circles, Al (measured with fast electrons) /7.38/; plusses, Al (measured with fast electrons)/7.36/; open circles, Si (measured with fast electrons) /7.39/; triangles, Be (measured with X-rays) /7.33/; crosses, Be (measured with X-rays) /7.32,35/

In Fig. 7.12 the values of $\omega(q^2)$ are collected which have been measured up to now. The general trend is easy to recognize. The points (•) /7.38/ which follow more or less the RPA curve (see Fig. 7.16) show a different behaviour.

We conclude from these observations that the collective excitation changes its character at $q \sim q_c$. This is supported by the observation that the linewidth of the Al 15 eV loss as a function of θ in the region around q_c has a strong increase, see Fig. 7.13. This indicates the onset of the decay of the collective oscillations into a single particle excitation if one passes the Landau edge.

Theoretical remarks. Under the conditions of a free electron gas and no electron-electron correlation we can follow the development of the excitation up to large q values and see the transition of the collective excitation into the single particle excitation. If the loss function $\mathrm{Im}\{-1/\varepsilon(\omega,q)\}$ using the Lindhard dielectric function together with the damping constant $\gamma = 1/\omega\tau$ /7.41/ is calculated for Al with

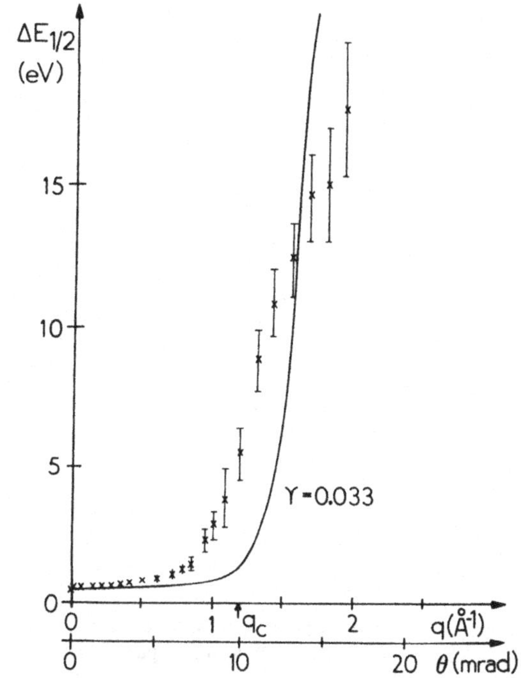

Fig. 7.13. Wave vector dependence of the loss width $\Delta E_{1/2}$ of the volume loss 15 eV Al. The solid line is calculated for the damping parameter $\gamma = 0.033$. $\gamma = 1/\omega\tau = 0.033$ is equivalent to $\Delta E_{1/2} = (1/30)\,\Delta E_p$ /7.36/, see also /7.40/

the observed $\gamma = 0.03$ one obtains Fig. 7.14. It displays Im$(-1/\varepsilon)$ at different q values up to 1.4 Å^{-1} just above $q_c = 1.3$ Å^{-1}. In the low-energy region a small peak appears due to the electron-hole excitation. It has a rather small value due to the strong screening of the plasma electrons against the incoming electron. Its intensity increases rapidly with q, due to the strong reduction of the screening and its maximum displaces to higher energies. At larger q values the single particle peak melts together with the plasmon and at $q \gtrsim q_c$ an asymmetric peak remains. This development can be seen by cutting, in Fig. 7.15, the plasmon and the single particle curve by a vertical line, so that Fig. 7.14 results for $q < q_c$. For experimental indication of these single particle losses see the end of this chapter.

To get the excitation probability of this process the loss function in Fig. 7.14 has to be divided by q^2; this leads to decreasing values of the probability with increasing q in the case of electrons as exterior perturbation. Since the loss function itself decreases at larger q values the loss intensity diminishes with increasing q quicker than q^{-2}, see Sect. 3.1.

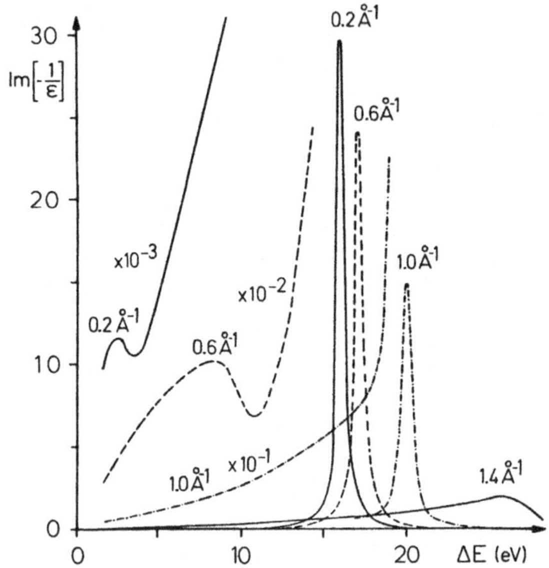

Fig. 7.14. Loss function of Al with the wave vector q as parameter and with a damping $\gamma = 0.03$ or $\Delta E_{1/2} = 0.5$ eV. The plasma loss and the weak single particle excitation are visible at low q values

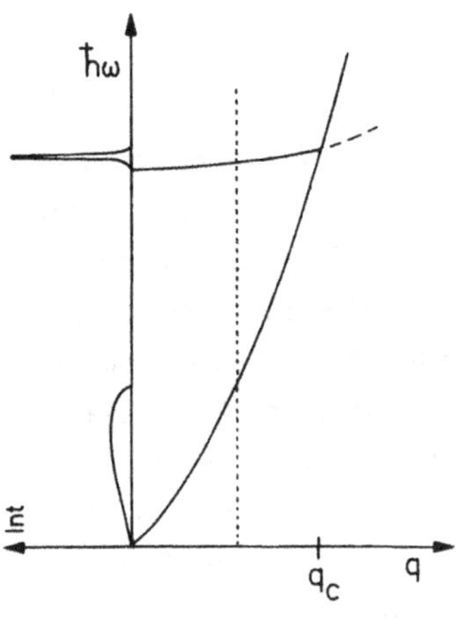

Fig. 7.15. The figure demonstrates the origin of the two loss peaks, the plasmon, and the single particle continuum at low energies for $q < q_c$ in Fig. 7.14

In the case of X-rays the intensity grows with $q^2 \cdot \mathrm{Im}(-1/\varepsilon)$ up to q_c, see (3.16). Calculations for beryllium show the increase of the maximum of the (differential) cross section up to about q_c and its decrease due to the decrease of the loss function for $q > q_c$ /7.41/; see also Fig. 3.7.

Plotting now the energy position of the loss maximum as a function of q one obtains a curve which shows a break in the slope in the region of q_c indicating the transition into the single particle excitation, as Fig. 7.16, curve A, shows. The difference of the slopes depends on the value of τ or $\gamma = 1/\omega\tau$; in Fig. 7.16 (curve A) γ has the value of 0.033, which follows from $\Delta E_{1/2} = 0.5$ eV for the Al 15 eV loss. A similar result has been obtained for the loss maximum of Be/7.41/.

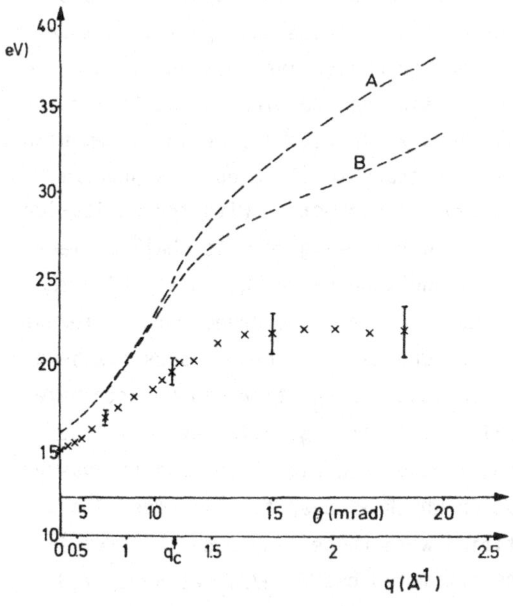

Fig. 7.16. Experimental dispersion relation of the Al 15 eV loss (x) /7.36/ and calculated data: A with Lindhard-Mermin and B with VASHISHTA-SINGWI /7.45/

The comparison of the observed with the calculated data shows a qualitative agreement insofar as a break of the slope is observed around $q \sim q_c$; the slope however especially that of the region $q > q_c$ in which the inelastic peak, not the plasmon, persists is steeper than the observed one. Variation of γ for Al within realistic values does not improve the situation /7.36/.

Since experiments with momentum transfer of $q > q_c$ probe the electron distribution around an electron within the screening length, exclusion principle and Coulomb interaction influence this distribution strongly.

If this electron-electron correlation shall be included into $\varepsilon(\omega,q)$, ε_{RPA} has to be corrected. Several attempts have been made /7.27,44-46/; however the calculated slope of $\omega(q^2)$ comes out essentially higher than the observed slopes /7.36,37/. Thus the disagreement persists.

Experiments with X-rays have explored the region between $\sim q_F$ and $\sim 2q_F$ /7.42/ where the inelastic peak has become rather broad. In Fig. 7.17 one remarks that the loss has a width of about 10 eV at $q/q_F = 1.5$ [q_F(Be) = 1.3 $Å^{-1}$; E_F(Be) = 13.8 eV]. Here at $q > 1.13\ q_F$ a new structure appears in the broad inelastic peak (of $\sim 3\ E_F$ halfwidth) which splits up into two humps, see Fig. 7.17. That of lower energy shows a weak or negative dispersion and disappears at about 2.1 q/q_F in the broader RPA-like peak. This structure is observed at different substances as Be, Al, graphite /7.42/, and Li /7.43,46a/. The authors ascribe this peak to a mode of a more or less ordered arrangement of the electron collective, not included in RPA. It can however hardly be regarded as a contribution of the plasmon excitation into the particle-hole continuum. Recent calculations /7.44/ seem to be in agreement with these experimental results. Further work is necessary to clear up this problem.

Excitation of the single particle continuum. In connection with the excitation of the single particle continuum, see above, the following results shall be mentioned: In the Al loss spectrum at $\Delta E < 15$ eV and at q values $q > 0$ at which the surface contribution has disappeared a series of peaks is observed with different dispersion behaviour /7.47,48/, see Fig. 7.18. One part of these losses has been associated with nonvertical, direct band transitions; the other can be attributed to the continuum excitation described in Fig. 7.14. In Fig. 7.18 the full line represents the low-energy edge of the continuum band, see Fig. 7.15, and the dashed line indicates the maximum of this excitation which is near to this edge. The losses near this dashed line can be identified with these single particle excitation peaks, its intensity being near the estimated one /7.47/, see Sect. 7.3.

7.3 Excitation of Bound Electrons with Momentum Transfer

In Sect. 7.1 we have treated the wave vector dependence of the collective oscillations of free electrons or of the intraband transitions. Here we discuss the q dependence of the excitation of bound electrons or of the interband transitions.

Electrons can excite crystal electrons by inelastic collisions transferring besides the energy ΔE a momentum $\hbar q = \hbar k_{el}\theta$. Direct nonvertical interband transitions are thus produced in which the electron takes up the (vertical) gap energy and in addition kinetic energy $(h^2/2m^*) \cdot q^2$ with m^* the effective mass. Momenta $\hbar q$ com-

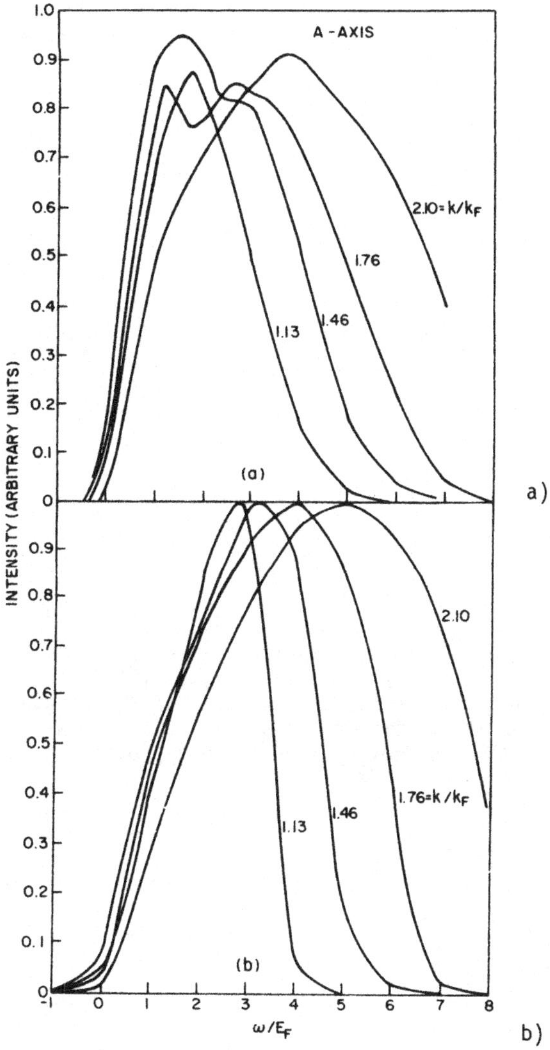

Fig. 7.17a,b. Plots of $(q^2/4\pi e^2)\,\mathrm{Im}(-1/\varepsilon)$ as a function of the loss energy with q as parameter (k is identical with q, k_F Fermi wave vector) (a) experimental results in Be. (b) Theoretical results obtained in RPA. Similar results were obtained in graphite and Al /7.42/

parable to and larger than the Brillouin zone can be transferred which is an important advantage of the ELS. The probability for such nonvertical transitions can be calculated, if $\varepsilon(\omega,q)$ is known which can be derived from the band structure in principle, see (2.18).

The first experimental evidence of these nonvertical transitions has been obtained by ZEPPENFELD on graphite monocrystals /7.12/ measuring the loss spectra as

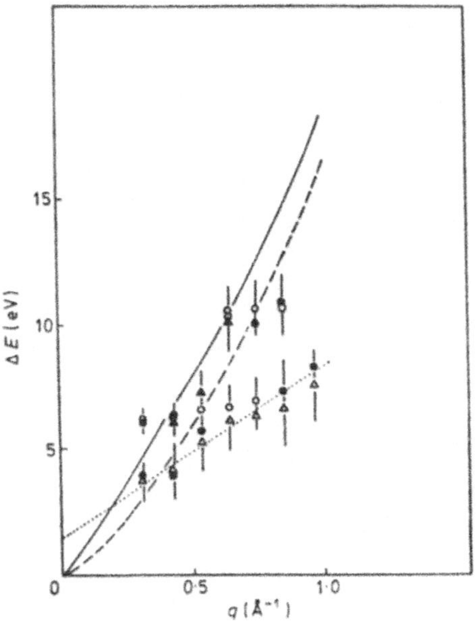

Fig. 7.18. The dispersion of the low-energy maxima in the loss spectrum of Al (epitaxially grown) (•) q ∥ [100], (o) q ∥ [110], (Δ) q ∥ [111], (▲) q ∥ [210]. The full line represents the low q boundary (curve (2) in Fig. 1.2) of the single particle continuum; the broken curve shows the maxima in the single particle excitation continuum calculated for $\gamma = 0.033$. The dotted line is interpreted as the dispersion of the interband peak at 1.5 eV /7.47/

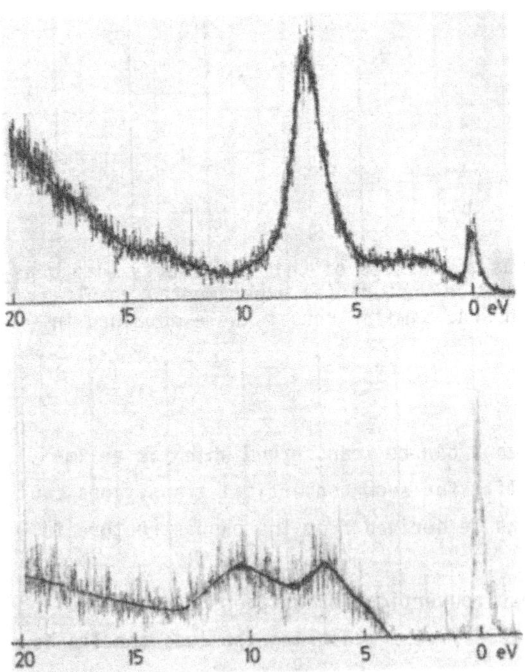

Fig. 7.19. Energy loss spectrum of graphite at θ_y = 0.8 mrad (upper spectrum) and at θ_y = 8 mrad (lower spectrum). θ_y or q_y, see Fig. 7.20 /7.12/

a function of the scattering angle θ. The change of the structure of the loss spectrum with θ could be explained by oblique transitions in the band structure of graphite.

Results of Measurements at Finite q Values

a) Dependence of the energy gap on q. The strong dependence of the loss spectrum of graphite on the value of q is demonstrated by Fig. 7.19. Here the angle θ or the vector q lies parallel to the direction q_y of the elementary cell in q space as it is drawn in Fig. 7.20. Besides a strong change of the structure of the spectrum, the onset of the loss intensity which starts in metals or rather good conductors as graphite from zero at q = 0 and grows linearly with q is displaced to larger q values, so that at higher q values the spectrum looks like that of an insulator, see Fig. 7.19, lower spectrum.

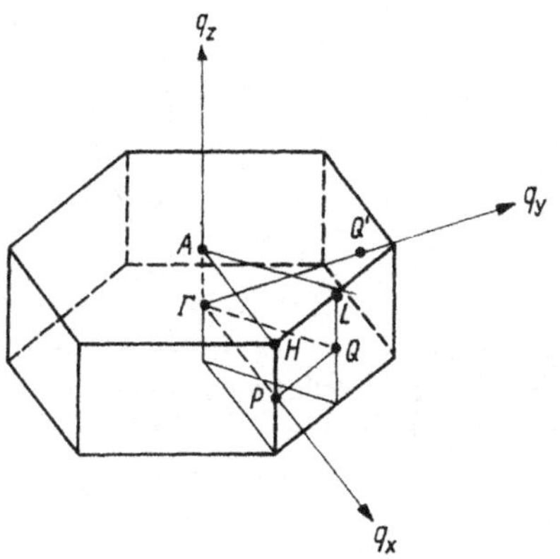

Fig. 7.20. Brillouin zone of graphite

The explanation of this behaviour is given for the graphite case in Fig. 7.21: here at the P_3^- point the Fermi energy passes the crossing point of two energy bands. Excitation of the electrons at the Fermi edge with a finite q transfer leads to a "gap", i.e., to a minimum energy which can be transmitted to the electron and which increases with q as Fig. 7.22 demonstrates. The observed dependence, see Fig. 7.22, shows further an anisotropy: it is different in $\overrightarrow{\Gamma P}$ and $\overrightarrow{\Gamma Q}$ direction which lie in the optically isotropic "a" plane of the crystal lattice) and along $\overline{\Gamma A}$; for higher q values the anisotropy of the Brillouin zone becomes apparent, since the dependency $E_g(q)$ splits for the two directions $\overrightarrow{\Gamma P}$ and $\overrightarrow{\Gamma Q}$ at $q \cong 0.5$ Å^{-1}.

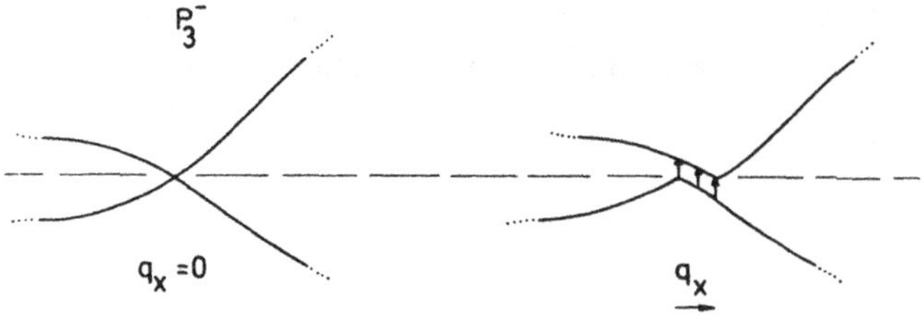

Fig. 7.21. Band structure of graphite at the P_3^- point at $q_x = 0$ and at $q_x > 0$. Dashed line: Fermi energy

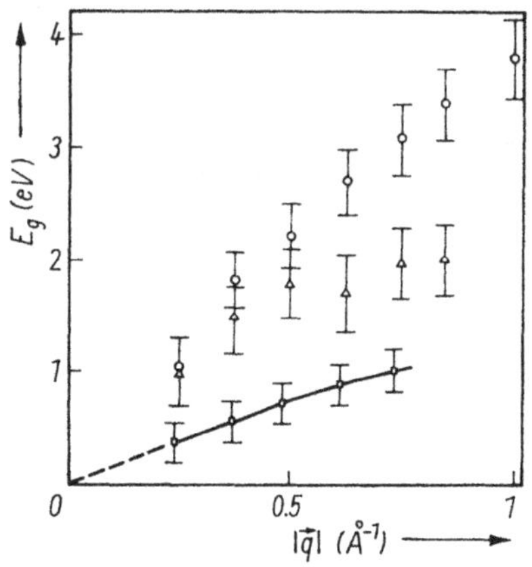

Fig. 7.22. Onset energy E_g as function of q in different crystal directions: (□) q ∥ $\overrightarrow{\Gamma A}$ (lowest curve), (Δ) q ∥ $\overrightarrow{\Gamma P}$, (o) q ∥ $\overrightarrow{\Gamma Q}$ /7.13/

These measurements allow one to compare the calculated and measured band energies; in this way it was found that the low-energy band has to be displaced to lower energies by about 1 eV /7.12,13/.

b) Dependence of the loss spectrum structure on q. Figure 7.23 shows an example of the q dependence of the loss function of graphite in the direction q_x in the plane perpendicular to the c axis, i.e., the q dependence of $\mathrm{Im}\{-1/\varepsilon(\omega,q_x)\}$.

From this loss function the dielectric function $\varepsilon_\perp(\omega,q)$ is derived via Kramers--Kronig analysis. The imaginary part $\varepsilon_{2\perp}(\omega,q_x)$ is reproduced in Fig. 7.24 and it shows how the "Drude" behaviour at very low q_x values disappears and a peak comes out which displaces with q_x. Table 7.6 shows how these peaks displace to higher values with increasing q_x. This dependence can be understood qualitatively from Fig. 7.21 where the high joint density between parallel energy bands produces a peak; this peak displaces to higher ΔE with higher q_x.

Table 7.6. Collection of the position in eV of the maxima and shoulders (s) of $Im(\varepsilon_{\perp x})$ and $Im(\varepsilon_{\perp y})$ at different $q_\perp$ values (in $Å^{-1}$) in graphite /7.24a/

q_x	Maxima of $\varepsilon_{2\perp}(\omega,q_x)$					
∿0		4.0 ± 0.2		12.5 ± 0.3	15.0 ± 0.4^s	
0.250	1.3 ± 0.2	4.4 ± 0.3		12.5 ± 0.3	15.1 ± 0.4^s	
0.375	2.1 ± 0.2	4.9 ± 0.3		12.6 ± 0.3	15.3 ± 0.4^s	
0.500	2.6 ± 0.3	5.1 ± 0.4		12.7 ± 0.4	15.7 ± 0.5^s	
0.625	3.0 ± 0.3			12.9 ± 0.4	16.5 ± 0.5^s	28.0 ± 0.6^s
0.750	3.4 ± 0.3			13.2 ± 0.4	16.7 ± 0.5^s	28.5 ± 0.6^s
0.875	3.6 ± 0.4		7.0 ± 0.5^s	13.4 ± 0.4	17.0 ± 0.6^s	29.0 ± 0.7^s
1.000	4.0 ± 0.4		7.2 ± 0.5^s	13.5 ± 0.5	18.5 ± 0.6^s	30.0 ± 0.7^s
q_y	Maxima of $\varepsilon_{2\perp}(\omega,q_y)$					
∿0		4.0 ± 0.2		12.5 ± 0.3	15.0 ± 0.4^s	
0.250	1.3 ± 0.2	4.4 ± 0.3		12.6 ± 0.3	15.1 ± 0.4^s	
0.375	2.1 ± 0.2	4.5 ± 0.3		13.0 ± 0.3	15.2 ± 0.4^s	
0.500	2.7 ± 0.3	4.7 ± 0.4	5.9 ± 0.5^s	13.3 ± 0.4	15.6 ± 0.5^s	
0.625	3.1 ± 0.3	4.8 ± 0.4	6.6 ± 0.5^s	13.9 ± 0.4	16.5 ± 0.5^s	
0.750	3.4 ± 0.3		7.4 ± 0.5^s	14.4 ± 0.4	17.0 ± 0.5^s	
0.875	3.7 ± 0.4		8.2 ± 0.5	15.4 ± 0.4	17.7 ± 0.6^s	
1.000	4.0 ± 0.4		8.9 ± 0.5	15.8 ± 0.5	18.8 ± 0.6^s	

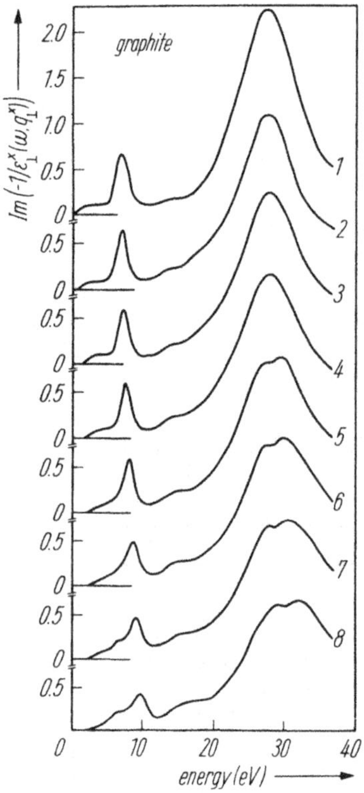

Fig. 7.23. The loss function of graphite in the direction $\overline{\Gamma P}$ with $q_\perp^x$ ($=q_x$) as parameter, see Fig. 7.20. The numbers mean the q_x values in $Å^{-1}$: 0; 0.25; 0.37; 0.50; 0.62; 0.75; 0.87; 1.00. $q(\overline{\Gamma P}) = 1.7$ $Å^{-1}$ /7.24a/

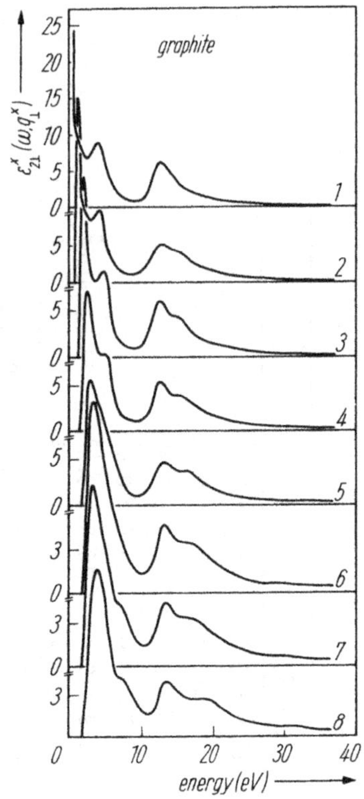

Fig. 7.24. The imaginary part of the dielectric function $\varepsilon_{2\perp}(\omega,q_x)$ in the direction $\overline{\Gamma P}$ /7.24a/

A quantitative calculation with the two-dimensional band structure approximation of BASSANI-PARAVICINI /7.49/ has been performed /7.12/. For this purpose the joint density of the nonvertical direct transitions has been calculated, neglecting however the q dependence of the matrix elements. Figure 7.25 shows the result of the calculations of $Im(\varepsilon_{\perp x})$ and $Im(\varepsilon_{\perp y})$ for the π band transitions of graphite. The σ band transitions ($\gtrsim$ 10 eV) have not yet been calculated. The agreement with the observed values of both curves is rather good regarding the approximations.

Similar observations of q dependent loss spectra have been reported on different substances such as BN /7.24a/, PbS, PbSe, PbTe /7.24b/, Al /7.47,48/, Ag /7.50/, and Si /7.39,51a/. Qualitative considerations of the nonvertical transitions in the band structure scheme can explain that peaks result at certain q values at which the energy bands are nearly parallel so that the joint density becomes large. The calculation of the joint density in three dimensions is rather cumbersome. In the case

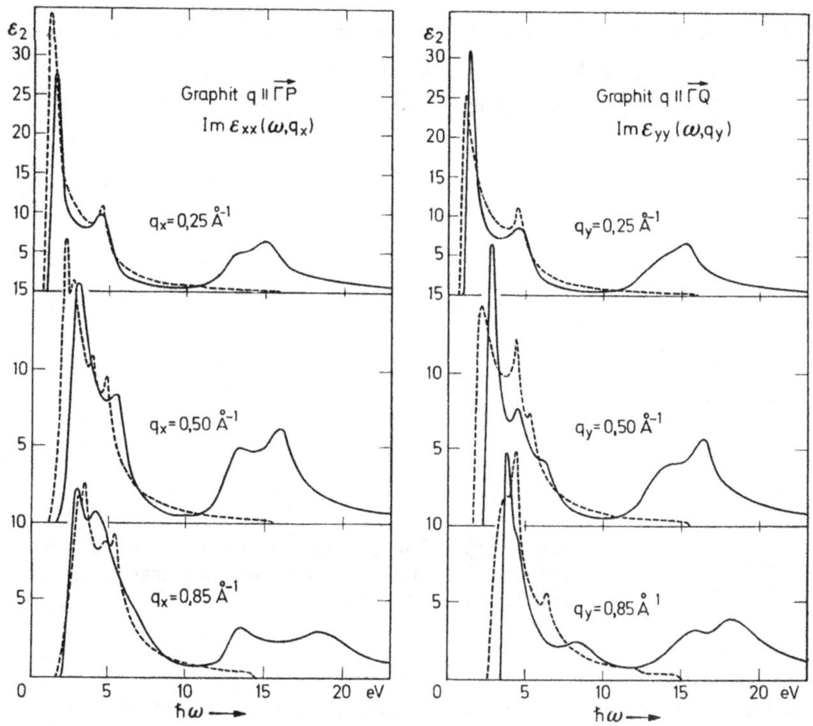

Fig. 7.25. The q dependence of $\varepsilon_{2\perp}(\omega,q)$ in the crystal directions $\overrightarrow{\Gamma P}$ and $\overrightarrow{\Gamma Q}$; experimental (full line), calculated (dashed) /7.12/

of Al calculations /7.51/ verify that the q dependent structure of the loss spectrum comes from interband transitions. Some of the loss maxima in Al are attributed to the excitation of electron-hole pairs, see the end of Sect. 7.2.

Dispersion of excitons. In a number of crystals excitons have been observed with ELS; the initial states of the electron are the valence band or deeper levels as, e.g., in KBr /7.52/ and in the Li halides near 60 eV /7.54/.

Solid Ne shows a very sharp exciton at 17.74 eV with $\Delta E_{1/2}$ = 90 meV, see Fig. 5.2; it represents the first term n = 1 of the $\Gamma_{1/2}$ exciton series. The peaks at 20.3 and 21.2 eV correspond to n = 2 and n = 3. The intensity of the 17.7 eV exciton allows one to measure its q dependence; it results in practically no shift of the peak position up to $q \cong 0.35$ Å^{-1} (50 keV electrons) /7.53/.

Similar experiments have been made on the excitons of LiF (13.50 eV) and NaF (10.90 eV); a shift of maximum 1 eV at $9 \cdot 10^{-3}$ rad (E_0 = 50 keV) and 0.5 eV at $8 \cdot 10^{-3}$ rad, respectively, has been observed /7.54/.

The q dependence of the π-π^* excitons of solid aromatics like anthracene, coronene, and p-terphenyl which lie in the energy region of about 6 eV, see Fig. 7.26, has been measured up to 5 mrad (E_0 = 50 keV); no shift is to be seen /7.55/.

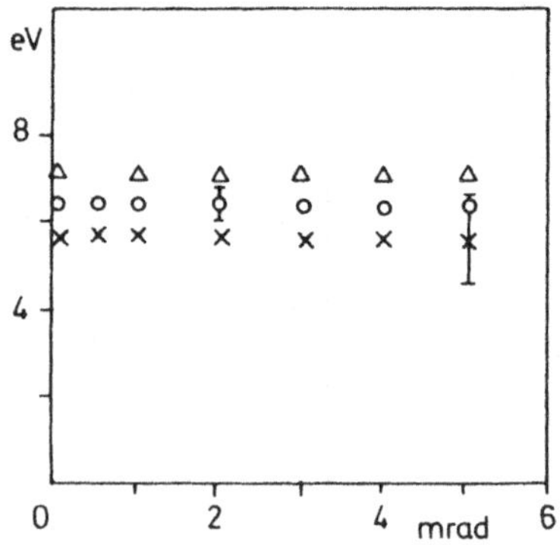

Fig. 7.26. The dispersion of excitons of the π-π^* excitations in coronene, anthracene, and p-terphenyle /7.55/

These results can be interpreted as follows: Band structure calculations for alkali halides result in an effective mass in the conduction band between 1/2 and 1 m_0, whereas the valence band has a rather small curvature so that the hole masses are rather large. With the figures given above for the 13.5 eV loss in LiF we obtain from the exciton equation

$$\Delta E_{ex}(q) = \Delta E(o) + \frac{\hbar^2}{2(m_{el} + m_h)} \cdot q^2 \tag{7.6}$$

a value of $(m_{el} + m_h) \cong 4$. This gives $m_h > 1$ in agreement with what has been derived from the band structure. We find a similar situation in rare gases and in aromatics.

7.4 Anisotropy of the Intensity of Volume Plasmon Losses

Experiments with monocrystalline films of Si, Ge, and epitaxially grown Ag films demonstrated a dependency of the volume loss intensity on the crystal orientation. The measurements were performed in transmission with 50 keV electrons in an arrangement sketched in Fig. 7.27. The intensity of the no loss peak and of the volume loss peak has been recorded as function of α by inclining the crystal against the electron beam. The result obtained on a thin Si monocrystal thinned by grinding and sputtering is shown in Fig. 7.28. The intensity of the no loss peak $[I(\Delta E = 0)]$

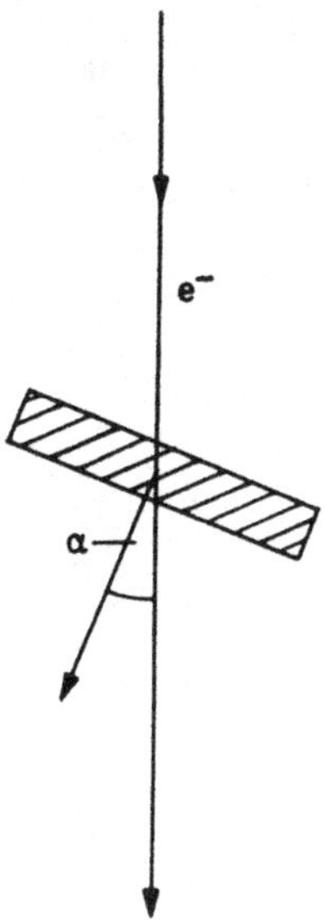

Fig. 7.27. Schematic arrangement to measure the intensity of the loss and of the no loss peak as a function of α. The intensity is measured in all these cases in the direction of the primary beam (θ = 0) /7.57/

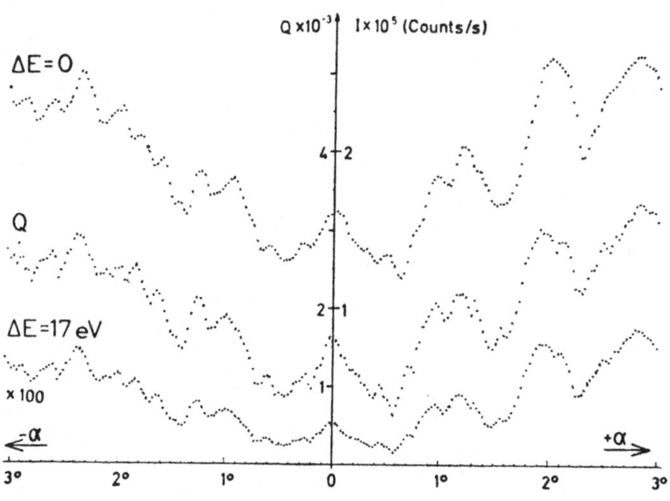

Fig. 7.28. The intensity of the no loss peak I(0) and of the Si volume loss peak I(ΔE) as a function of the angle α, see Fig. 7.27. From both the quotient Q = I(ΔE)/I(0) is derived. Thickness of the crystal about 1000 Å /7.56/

fluctuates with α, a behaviour which can be understood in the framework of the dynamical electron diffraction concept. One observes the same for the inelastic peak intensity $I(\Delta E = 16.9\ eV)$, since a proportionality of $I(\Delta E)$ and $I(0)$ is expected. Not yet understood is the result that $Q = I(\Delta E)/I(0)$ is not a constant but varies in correlation with $I(0)$ /7.56,57/.

8. Core Excitations. Application to Microanalysis

The region of energy losses which has been described in the foregoing chapters comprises the collective excitations of the loosely bound electrons as the valence electrons of a metal and interband excitations of the order of 10 eV. In addition to these losses RUTHEMANN /8.1/ observed in his transmission experiments high-energy losses of 100 eV and more due to the excitation of deeper atomic levels (core levels) known from X-ray absorption. Figure 8.1 shows the energy loss of 291.2 eV at the K edge of the C atom. These losses can be described as practically single electron excitations in contrast to the plasmons; collective contributions can be neglected in a first approximation.

The interest in these losses lies a) in the possibility to identify small quantities of a substance in an electron microscope by these characteristic excitations; b) in the remarkable fine structure of their edges in the solid.

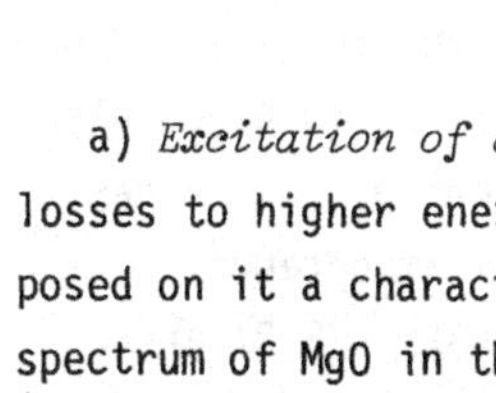

Fig. 8.1. Energy loss of 291.2 eV produced in a film containing carbon by the excitation of the K shell of the C atom. Electron energy 7.5 keV /8.1/

a) *Excitation of deep levels.* If one extends the study of the electron energy losses to higher energies, one observes a continuously decreasing background, superposed on it a characteristic weak loss structure. Figure 8.2 displays such a loss spectrum of MgO in the high-energy region. Here the K shell of oxygen (532 eV) and that of magnesium (1305 eV) are to be seen; the loss peaks lie on a rapidly decreasing background.

The interest of these experiments lies in the fact that these deep level losses originating from atomic levels are characteristic for the irradiated specimen and allow one thus to derive its chemical composition. Since the electrons can be fo-

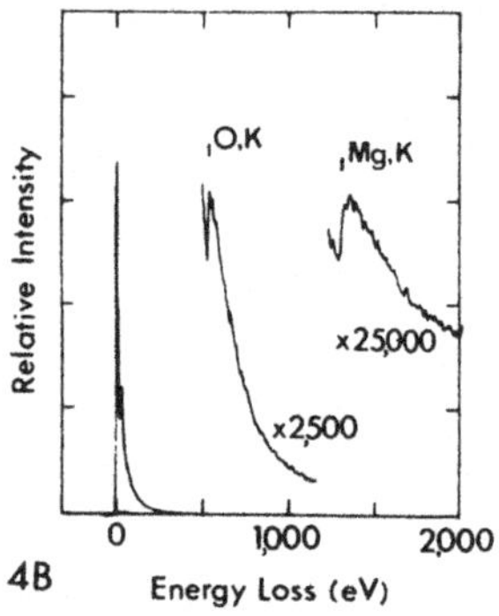

4B

Fig. 8.2. Loss spectrum of MgO in the energy region of about 1000 eV /8.2/

cussed by lenses to very small areas the microanalysis of rather small regions of a specimen is possible.

The plasmons are less suitable for such an identification since here do not exist many elements with pronounced volume losses. A mixture of two metals with volume losses $\hbar\omega_A$ and $\hbar\omega_B$ does not show two volume peaks of different heights corresponding to their quantity but it will certainly produce one plasmon peak $\hbar\omega_{AB}$ given by the individual alloy; it is thus not possible to derive from this mean peak, displaced against the original peaks $\hbar\omega_A$ and $\hbar\omega_B$, the components and their quantities.

A further limitation to use the volume plasmon for a quantitative analysis comes from the fact that the excitation probability of a volume plasmon is only proportional to the crystal thickness, see (3.3), as long as the surface probability does not contribute to the intensity at $\varepsilon = 0$, see (10.49). For thicknesses $< \sim 100$ Å surface contributions are not negligible; they depend on surface contaminations in general not known.

The well-known X-ray fluorescence method uses fast electrons too to excite atoms, but it measures the emitted X-ray intensity produced by the deexcitation of the atoms. Here in ELS the energy loss of the fast electrons is observed. This method has advantages especially for the detection of light elements which have a very low fluorescence yield since most of the excitation energy is transferred to Auger electrons, whereas in ELS every electron having lost its excitation energy can be detected. The application to biological specimen is thus of special importance.

This procedure has been proposed and realised by HILLIER and BAKER /8.3/ and developed further at different places /8.4-7/.

Important progress has been made by CREWE by the introduction of the field emission cathode requiring ultrahigh vacuum conditions better than some 10^{-10} Torr /8.8,9/. This improvement had two consequences: 1) the diameter of the irradiated specimen can be reduced so that values of the electron beam diameter of about 5 Å have been achieved; 2) the low pressure in the analyzer diminishes the contamination rate produced in the irradiated area enormously. In addition the energy width of the primary beam could be reduced to about 200 meV.

The success of these experiments encourages one to visualise the possibility of a quantitative microanalysis which uses the absolute loss intensity to obtain the mass of the irradiated crystal. Up to now this determination has an accuracy of about 20 %, see the detailed discussion in /8.10,11/. As Fig. 8.2 and Fig. 8.3b show one needs the height of the signal above the background which is not easy to fix. The latter stems from the high energy tail of the excitation at lower energies. Multiple losses of lower energies are certainly effective too. An improvement of the signal to background ratio is possible: The angular distribution of the inelastically scattered electrons is rather concentrated in the forward direction as has been explained in Chap. 3: the main intensity of this volume excitation decreasing with $(\theta^2 + \theta_{\Delta E}^{\ 2})^{-1}$ lies inside an angle of about $\theta_{\Delta E} = \Delta E/2E_0$. This forward concentration is an essential advantage of the electron losses compared to the X-rays which are emitted into a solid angle of 4π and cannot be collected as efficiently in a detector. It is therefore important in ELS to adapt the angular aperture of the detector to the distribution of the scattered electrons, so that on does not accept too much of the background and does not lose too much of the signal intensity.

As an example of the possibility of the procedure the dark-field image of a small crystal of 60 Å in size of ferritin $(FeO \cdot OH)_8\ FeO \cdot OPO_3H_2$ is shown in Fig. 8.3a. The arrow indicates the crystal which has been irraditated to get the loss spectrum of Fig. 8.3b. Here the loss intensity I divided by the intensity of the no loss peak I_0 is drawn in a log-log plot to derive the background intensity; it is evident that the oxygen K edge and the iron L_{23} edge is well detectable and allows quantitative evaluation. One can estimate that the crystal of 60 x 60 x 60 Å has a weight of some 10^{-19} gr, which demonstrates the sensitivity of this method. For more details see /8.10,11/.

Using the possibilities of the high-voltage electron spectroscopy the application promises to get a higher sensitivity: the smaller $\Delta E/2E_0$ values reduce the acceptance angle and maximize the signal to background quotient and further the background shall decrease with increasing energy (E_0) of the exciting electrons /8.13/.

If one is interested in the chemical composition of the specimen at other different positions, the focussed electron beam can be scanned over the film and the data are stored electronically. Further the specimen can be looked at through different energy loss windows between $\delta E + E$ and E, e.g., in the "light" of the K edge by special energy filters in the electron microscope.

b) *Structure near the absorption edges.* If an electron is excited in a deep level of an atom of a solid, one observes a fine structure of the absorption edge on the high-energy side of X-rays as well as of the energy losses of the inelastically scattered electrons in the energy region of 10^1 to 10^2 eV. This comes from a modu-

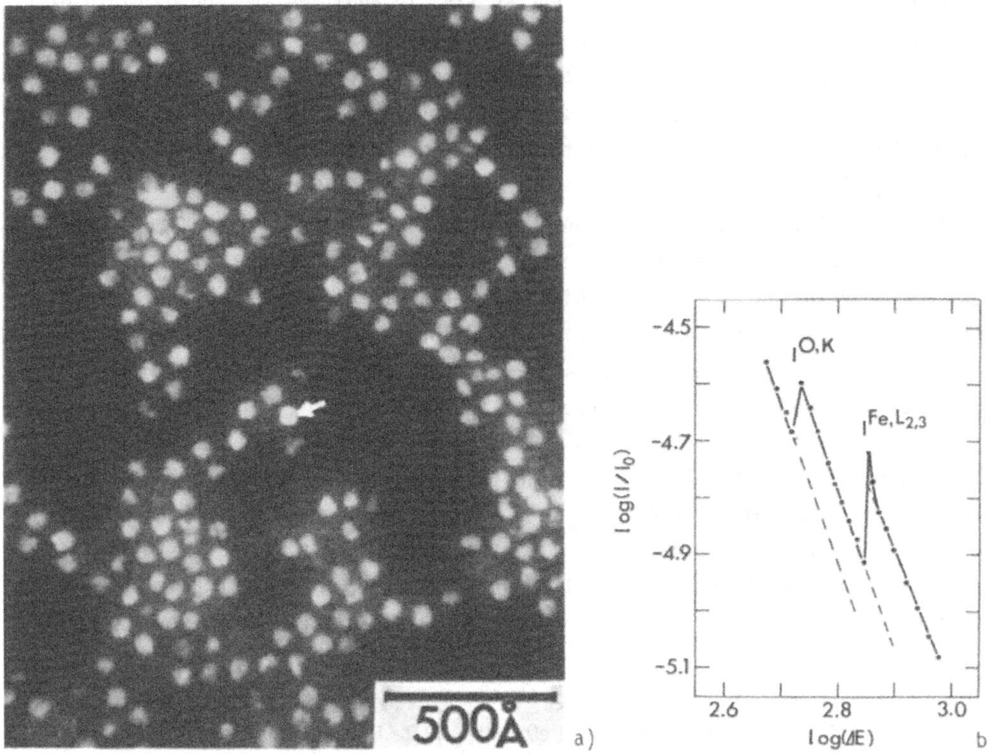

Fig. 8.3a,b. A dark-field image of ferritin particles (a) and its energy loss spectrum on a log-log plot (b). The background falls off as ΔE^{-n} with n = 2.5-3. ΔE means the energy loss and I_0 the intensity of the no loss peak /8.2/

lation of the transition probability from the occupied core levels into those above the Fermi level which is strongly influenced by the atomic short range order in the neighbourhood of the excited atom. This fine structure of the absorption edge of X-rays has been observed and interpreted already a long time ago /8.14/. This phenomenon called EXAFS (extended X-ray absorption fine structure) is observed as structure in the electron energy loss spectrum or in the absorption coefficient of X-rays above the absorption edge. Figure 8.4 /8.15/ shows this effect at the L edge of Al which extends over nearly 10^2 eV. The situation in this case is as follows: An energy loss of the incoming electron takes place if the excited electron can occupy the first free level in the conduction band just above the Fermi level, see Fig. 8.5. This onset will take place in aluminum at 72.72 eV (L III) and 73.15 eV (L II) in agreement with X-ray observations. However the loss intensity does not fade out continuously with increasing loss energy, but it shows a characteristic structure extending to rather high values ($\sim$ 130 eV), different for the various

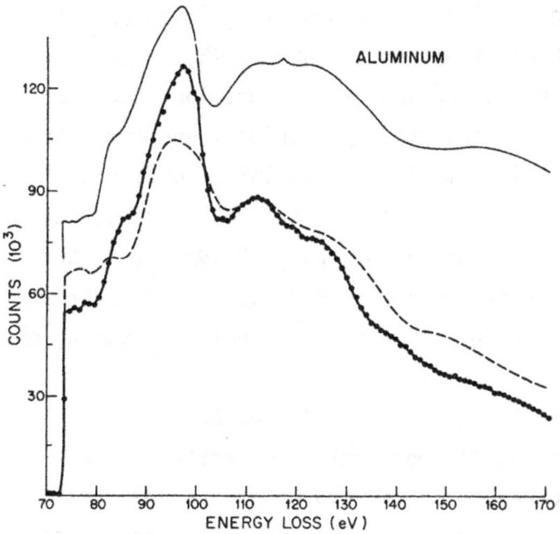

Fig. 8.4. L_{II}. L_{III} absorption edge of aluminum. Data derived from electron energy losses (solid circles) and from X-ray absorption (solid line) compared with calculated values (dashed line) (the solid line is displaced upward). The absorption data of X-rays can be compared with the loss function of electrons, since $\varepsilon'' \ll 1$ and $\varepsilon' \sim 1$, so that $\mathrm{Im}(-1/\varepsilon) \sim \varepsilon''$ /8.15/

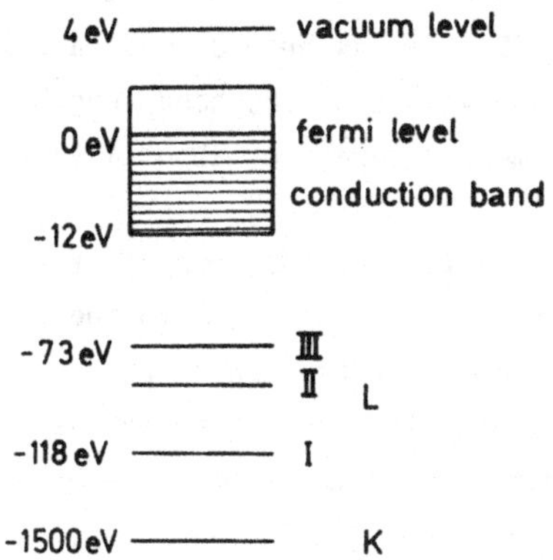

Fig. 8.5. Energy levels in Al

substances and different for the liquid and the crystal states. For comparison: the absorption of photons of the synchrotron radiation in Fig. 8.4 shows similar behaviour. This phenomenon is explained by a rather simple picture /8.16,18/: The excited electron waves leave the atom, are backscattered partially by the surrounding atoms, and interfere at the position of the excited atom. Thus the transition probability of the electron from the initial level to the outer levels is modulated depending on the energy of the outgoing electrons and on the arrangement of the surrounding atoms. In this way the band structure of the solid determines the structure of the absorption edge. The theoretical dependence of the loss intensity as function of the energy loss in Al is reproduced in Fig. 8.4 and shows good agreement with the experimental result obtained with electrons or photons /8.15/.

The intensity variation is a result of phase and amplitude summation at different energies of the scattered electrons. It can be inverted by Fourier transform into a radial distribution function which contains information of the interatomic distances and the number of atoms in the coordination shells as has been demonstrated on crystalline and amorphous Ge /8.17/ and recently on Graphite /8.19/. It is characteristic of this method that the atoms in the bulk can be discerned by their absorption edges which allows one to separate in a complex substance the near neighbour coordination around different atoms.

More details of the theoretical considerations are found in /8.18/.

For the evaluation of the EXAFS spectrum it is important to separate the interference pattern from the monotonically decreasing background which has then to be corrected for multiple scattering etc.. In the further steps one has to calculate among others the phase of the electron wave which leaves the excited atom and comes back after being reflected at the surrounding atoms. The phase shift is determined by the potential in which the electron moves on this way and by its energy which varies between some eV and 10^2 eV. These values together with the amplitude of the backscattered wave are known from theoretical work.

An idea of the accuracy gives the comparison of the bond lengths to nearest neighbours obtained from X-ray diffraction and from EXAFS. It has been demonstrated, e.g., for Cu and some of its compounds that the absolute values of the bond length agree within 0.02 Å /8.20/. Relative measurements are more accurate by about a factor of 10 /8.21/.

In anisotropic crystals orientation dependence is expected and has been verified by experiments on Zn monocrystals: if the polarisation vector is parallel to the c or the a axis different EXAFS spectra are observed. Since the ratio c/a in Zn is larger than the normal value (1.633), the nearest neighbour in the c direction is 0.222 Å farer away than in the a direction as X-ray diffraction has shown. EXAFS measurements give a value of 0.225 ± 0.005 Å in good agreement /8.22/.

Different other applications of EXAFS have been reported: The adsorption of Br_2 (0.2 monolayer) on Graphite surfaces has been observed with different polarisation of the X-rays; the evaluation shows that Br_2 is adsorbed with one atom fixed above the basal-plane hexagonal site; the axis of the molecule can move freely /8.23/.

The EXAFS of Ag I (Ag K shell) has been applied to determine the position and dynamics of the Ag ions in superionic Ag I (α phase) which has an ionic conductivity by four orders higher than the ordinary Ag I (β phase) (transformation temperature 147 $^{\circ}$C). The distribution function of the position of the Ag ions could be determined as well as a comparison of the residence time of the ion in its site with the flight time between different sites was possible; the latter comes out three times smaller /8.24/.

The information which is hidden in the details of the extended fine structure of the edges is not yet finally explored and a lot of work is going on concerning this subject.

Another type of structure of the edge exists at the threshold of the core excitation of metals. An electronic excitation from a deep level to a state near the Fermi level can go on as a single electron transition, so that the excited electron alone is involved. It is possible however that this excitation is accompanied by a rearrangement of the conduction electrons near the core hole. In this case the shape of the absorption edge is changed. It is not yet cleared up whether observed deviations from the single particle behaviour are to be interpreted as such a many body effect.

9. Energy Losses by Excitation of Cerenkov Radiation and Guided Light Modes

9.1 Cerenkov Radiation

If the velocity of light in a medium with the refractive index $n(\omega) = \sqrt{\varepsilon'(\omega)}$ is smaller than the velocity v of electrons which pass the substance: $c/n(\omega) < v$ or $1 < \varepsilon'\beta^2$, photons of the energy $\hbar\omega$ are emitted. The electron loses this energy and is deflected by an angle θ of roughly $\lambda_{el}/\lambda_{Ph} \sim 10^{-5}$ rad, $\lambda_{el} = 0.05$ Å (50 keV) and $\lambda_{Ph} = 5000$ Å (2.5 eV), which follows from $\hbar q = \hbar k_{Ph} n$ with $k_{Ph} = 2\pi/\lambda_{Ph}$ and $q = k_{el} \cdot \theta$.

The formula (3.3) for the plasmon excitation probability calculated without retardation does not contain these energy losses. The relativistic correction of the loss probability has been calculated by KRÖGER /9.1/. For our discussion of this rather lengthy correction it is sufficient to discuss the following term which is important for larger film thickness:

$$\frac{\partial^3 W}{\partial \Delta E \partial^2 \Omega} = D \cdot \frac{e^2}{\pi^2 \hbar^2 v^2} \cdot \mathrm{Im}(-1/\varepsilon) \frac{\theta^2 + \theta_{\Delta E}^2 \left\{(1-\varepsilon'\beta^2)^2 + (\varepsilon''\beta^2)^2\right\}}{\left\{\theta^2 + \theta_{\Delta E}^2 (1-\varepsilon'\beta^2)\right\}^2 + (\theta_{\Delta E}^2 \varepsilon''\beta^2)^2} \tag{9.1}$$

with $\beta = v/c$.

An important feature of this equation is that the denominator of the last term has a minimum leading to a strong additional maximum of the loss probability. Neglecting ε'' we have

$$\theta_{max} = \theta_{\Delta E}(\varepsilon'\beta^2 - 1) \tag{9.2}$$

with $\theta_{\Delta E} = \Delta E/2E_0$ (3.10) and $\varepsilon'\beta^2 > 1$ (Cerenkov condition). The maximum appears at small angles, $\theta_{max} \sim 10^{-5}$ rad, and has a rather high peak value; its position changes with the frequency ω of the emitted light as Fig. 9.1 demonstrates. Figure 9.1 shows further that no peak exists in the nonretarded case (dashed line). For $\theta > \theta_{max}$ the retardation influence disappears rapidly.

Folding the loss probability of (9.1) with the angular intensity distribution of the primary electrons which may have a halfwidth of about some 10^{-4} rad, one inte-

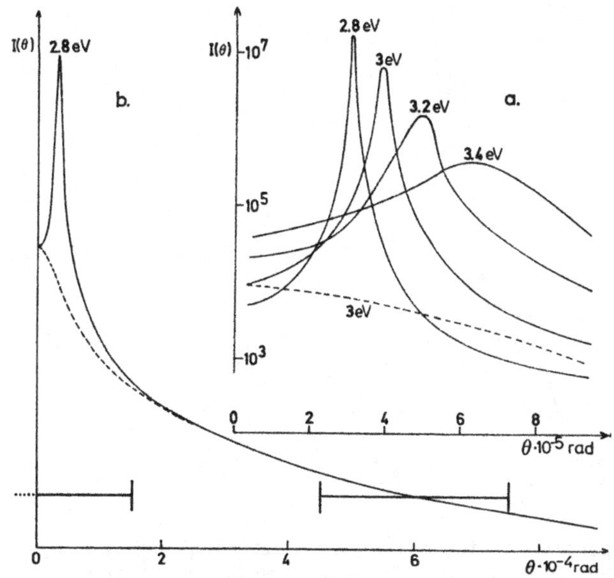

Fig. 9.1. Calculated loss intensity I(θ) of the volume contribution as function of the scattering angle θ. GaP, film thickness 2100 Å, E_0 = 40 keV. Full line: unfolded data; dashed line: nonretarded data. (a) (inset) for different photon energies (b) for 2.8 eV. The bars indicate the angular width of the primary electrons /9.5/

grates over the different peaks of Fig. 9.1a. A well-defined radiation loss which should be observable results.

Such loss peaks have been observed first in thin Si films at an energy of 2.9 eV /9.2/. Further experiments on other substances like Ge, III-V compounds, diamond, etc., which have rather high ε' values in the visible region (2-4 eV) showed similar effects /9.4,5/. The position as well as its absolute intensity has been calculated with the relativistic formula (9.1) using the known dielectric function /9.3/ and is in good agreement with the observations /9.4-6/. As an example the observed and calculated loss intensities are compared for GaP, see Fig. 9.2. For large scattering angles the experimental and calculated curve approaches the nonrelativistic dependence.

Instead of detecting the radiation by the energy loss of the electrons one can also measure the emitted photons. For thin films the radiation is in general caught in the film of high refractive index because of total reflection at the surfaces. However, if the normal of the thin film is inclined with respect to the electron beam the photons can leave the foil /9.7/.

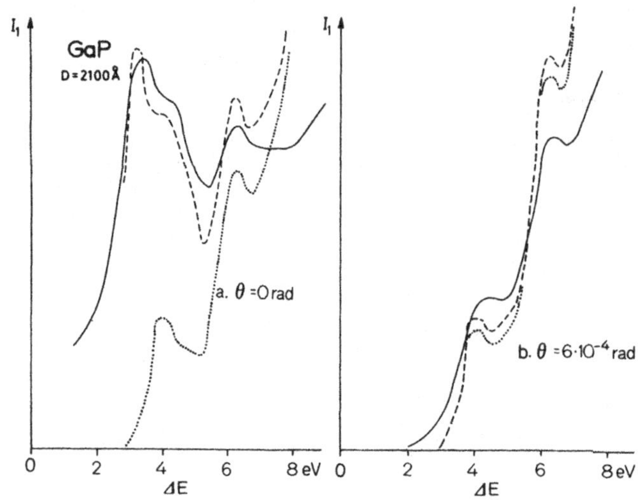

Fig. 9.2. Electron energy loss spectrum at two different scattering angles θ in GaP. Solid line: experimental curve. Dashed line: calculated with retardation. Dotted line: calculated without retardation. The "radiation loss" at about 3 eV in (a) disappears at higher scattering angles, curve (b). E_0 = 50 keV /9.5/

9.2 Excitation of Guided Light Modes by Electrons

Guided light modes can be excited in a thin dielectric film by electrons. The dispersion of the modes is displayed in Fig. 9.3. Theoretically these modes are obtained from the terms of the complete relativistic excitation probability in a film of thickness D /9.1/ containing the expression $1/L^+$ and $1/L^-$. If the denominators of $L^\pm$ [the dispersion relations, see (10.6)] have small values, maxima in the loss intensity appear. To describe the experimental results discussed below we specialize these equations for nonradiative light modes ($\varepsilon' > 1$), which give the curves of Fig. 9.4, lying between the light lines $\omega/q = c$ and $\omega/q = c/\sqrt{\varepsilon'}$. The observation of these light modes by ELS has been possible by using small angular width ($< 10^{-5}$ rad) of the primary electron beam, so that the losses lying in the region between the

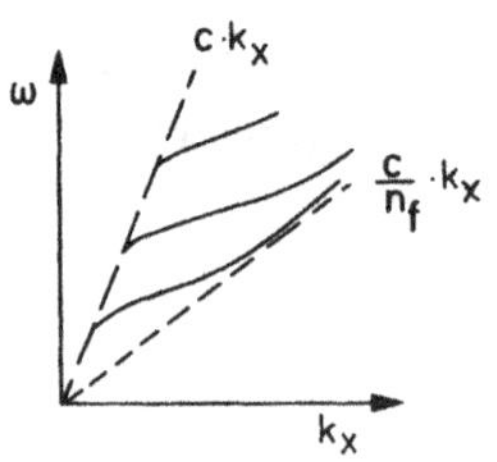

Fig. 9.3. Dispersion relation of guided light modes $\omega(k_x)$. In Fig. 10.10 and 10.11 we have drawn as abscissa $\theta = \arcsin k_x/(\omega/c)$ instead of k_x.($n_f = \sqrt{\varepsilon'}$)

light lines c and $c/\sqrt{\varepsilon^T}$ are detectable. Such modes have been photographed in graphite and Al_2O_3 /9.8,9/. The results of graphite are discussed in detail in /9.10/ and have been found in good agreement with the observed values.

Energy losses by transition radiation. Experiments with high-energy resolution have been performed in the infrared spectral region. The observed energy losses are attributed to the production of transition radiation /9.11/.

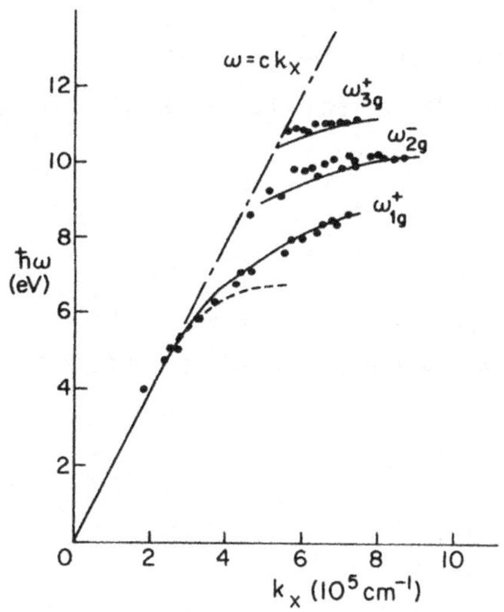

Fig. 9.4. Dispersion of three guided light modes in graphite. The solid curves are calculated values taken from optical data, the points are measured with ELS /9.9/

10. Surface Excitations

10.1 Surface Plasmons

Collective excitations of electrons exist not only in the volume of a plasma but also in its boundary /10.1/. Longitudinal waves of the surface charge density are possible which run along the surface as a polarisation wave

$$\sigma(x,t) = \Delta\sigma \exp i(\omega t - k_x x) \tag{10.1}$$

as RITCHIE has shown /10.2/. The charges displace parallel to the wave vector k_x which lies in the boundary. Figure 10.1 displays the situation of a p polarised

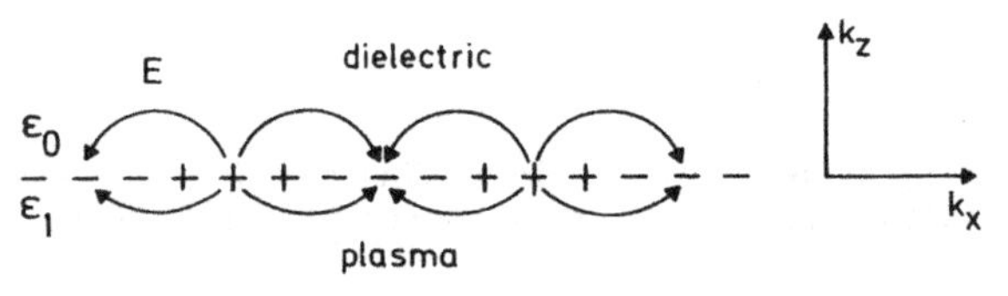

Fig. 10.1. Field configuration of a p polarised surface plasmon on a plasmon boundary. The charge oscillations are coupled with electromagnetic fields

surface wave. The displaced charges produce an electric field with components parallel and perpendicular to the boundary. The magnetic field lies perpendicular to this plane and has only components H_y. This field configuration propagates with the phase velocity ω/k_x.

The eigenfrequency ω of these waves is connected with k_x by an implicit dispersion relation which can be determined by starting from Maxwell's equations. We obtain for a film of thickness $d^{+)}$ and a frequency-dependent dielectric function

---

+) In this chapter the thickness has been named d instead of D.

$\varepsilon_1(\omega)$ covered on both sides by different media with dielectric constants ε_o and ε_2 (asymmetric case, see Fig. 10.2) for a p polarised wave the implicit dispersion relation

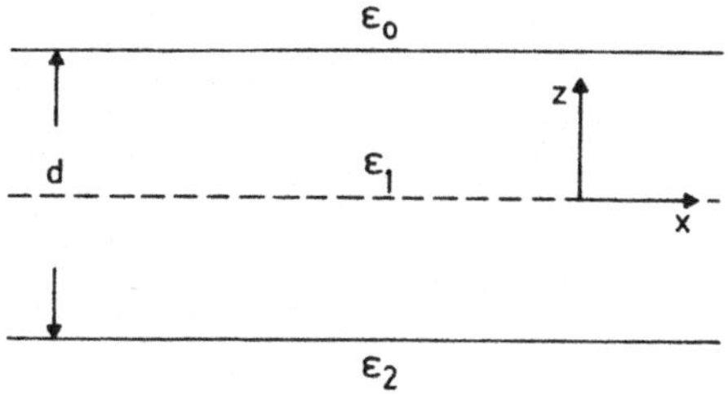

Fig. 10.2. A plane parallel plasma film with a dielectric function $\varepsilon_1(\omega)$ covered on both sides with media of different dielectric values ε_o and ε_2

$$\begin{aligned} L = & (\varepsilon_1 k_{zo} + \varepsilon_o k_{z1})(\varepsilon_2 k_{z1} + \varepsilon_1 k_{z2}) \\ & + (\varepsilon_1 k_{zo} - \varepsilon_o k_{z1})(\varepsilon_2 k_{z1} - \varepsilon_1 k_{z2})\, e^{2ik_{z1}d} = 0 \end{aligned} \tag{10.2}$$

and for a s polarised wave

$$(k_{zo} + k_{z1})(k_{z1} + k_{z2}) + (k_{zo} - k_{z1})(k_{z1} - k_{z2})\, e^{2ik_{z1}d} = 0. \tag{10.3}$$

k_{zi} and k_x are the components of the wave vector $\sqrt{\varepsilon_i}\,\omega/c$ with

$$k_x^2 + k_{zi}^2 = \varepsilon_i(\omega/c)^2. \tag{10.4}$$

The character of the solutions or the behaviour of the field outside the slab, e.g., of the component

$$E_z \sim \exp i(k_x x + k_{zi} z - \omega t) \tag{10.5}$$

is given by k_{zi}, i = 0 or 2.

a) Nonradiative modes: if k_{zi} outside the slab is imaginary, E_z decays exponentially with z and we speak of nonradiative modes. In this case we have $k_x > \omega/c$ ($\varepsilon_o = 1$), see (10.4), or the dispersion relation lies to the right of the light line, see Fig. 10.3.

b) Radiative modes: if k_{zi} outside the slab is real, the solution has an oscillatory character or the mode is a radiative one. This means $k_x < \omega/c$ or the dis-

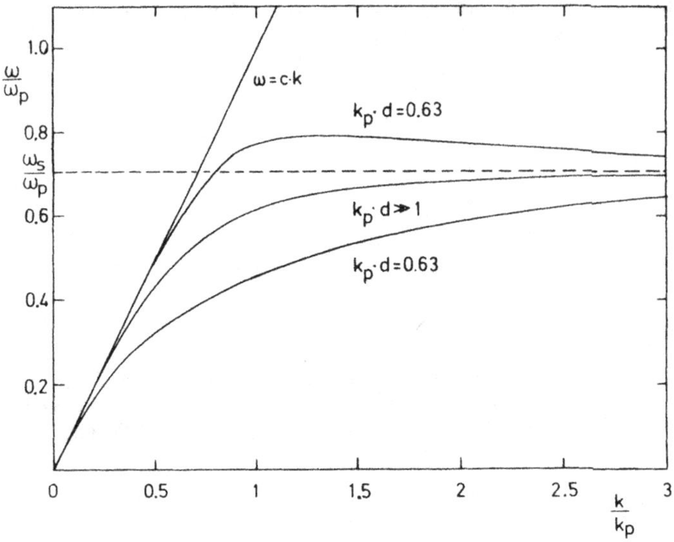

Fig. 10.3. Dispersion relation of nonradiative plasmons in a free electron gas for a thick film $k_pd >> 1$ and for a thin film $k_pd = 0.63$ ($k_p = \omega_p/c$), light line: $\omega = ck$

persion relation lies to the left of the light line. In the following we discuss these modes for a film of thickness d.

Nonradiative modes: k_{zo} and k_{zi} are imaginary. For the symmetric case ($\varepsilon_o = \varepsilon_2$) and the p polarised wave we obtain

$$L^+ = \varepsilon_1 k_{zo} + \varepsilon_o k_{z1} \tanh \frac{1}{2i} k_{z1} d = 0 \tag{10.6}$$

$$L^- = \varepsilon_1 k_{zo} + \varepsilon_o k_{z1} \coth \frac{1}{2i} k_{z1} d = 0. \tag{10.6a}$$

S polarised surface oscillations do not exist since E_y is continuous and thus no surface charge is produced.

We discuss (10.6) for different values of k_x. For *large* k_x values ($k_{zi} = i\,k_x$) (10.6) can be written

$$\frac{\varepsilon_o+\varepsilon_1}{\varepsilon_o-\varepsilon_1} = \mp e^{-k_x d} \tag{10.7}$$

which gives for $k_xd \to \infty$ the important relation

$$\varepsilon_o = -\,\varepsilon_1(\omega). \tag{10.8}$$

This (implicit) dispersion relation, neglecting retardation, can also be obtained as follows: in the interface between two media with the dielectric functions ε_0 and ε_1 the normal components of the electric displacement D_n have to be continuous or

$$\varepsilon_0 E_0^n(0) = \varepsilon_1 E_1^n(0).$$

In the interface containing the polarisation charges one has due to symmetry $E_0^n(0) = -E_1^n(0)$ (no external fields), so that $\varepsilon_0 = -\varepsilon_1$ /10.2a/.

In the case of a free electron gas we obtain for the eigenfrequency

$$\omega^{\pm} = \frac{\omega_p}{\sqrt{1+\varepsilon_0}} \sqrt{1 \pm e^{-k_x d}} \tag{10.9a}$$

which turns over for $k_x d \to \infty$ to

$$\omega_s = \frac{\omega_p}{\sqrt{1+\varepsilon_0}} \,. \tag{10.9b}$$

For *small* k_x values the dispersion relation (10.6) is described by

$$k_x = \frac{\omega}{c} \sqrt{\frac{\varepsilon_0 \varepsilon_1}{\varepsilon_0 + \varepsilon_1}} \tag{10.10}$$

which approaches the light line

$$k_x = \sqrt{\varepsilon_0} \cdot \omega/c \tag{10.11}$$

for $\varepsilon_1 \to -\infty$.

We see that the momentum $\hbar k_x$ of a surface plasmon, the quantum $\hbar\omega$ of the plasma oscillation, is larger than that of a photon of the same frequency by

$$\hbar(k_x - k_\ell) = \frac{\hbar\omega}{c} \sqrt{\varepsilon_0} \, \frac{\varepsilon_0}{2|\varepsilon_1|} \,. \tag{10.12}$$

In the intermediate k_x region $-\infty < \varepsilon_1 < -\varepsilon_0$ the dependence $\omega^{\pm}(k_x)$ is shown in Fig. 10.3; the splitting decreases with $k_p d$.

The existence of two solutions $\omega^{\pm}$ in the case of a thin film with symmetric coating ($\varepsilon_0 = \varepsilon_2$) results from the coupling of the two identical SP's (same frequency for given k_x) at both interfaces of the film which leads to a splitting: ω^+ describes the SP of higher frequency since charges of opposite sign lie on both sides of the film at a given point x whereas ω^- is smaller than ω^+ since charges of equal sign reduce the eigenfrequency. The electric field of the ω^- oscillation has a configuration symmetric with respect to $z = 0$ whereas it is asymmetric for the ω^+ oscillation.

For $d \to \infty$, the case of one boundary only, (10.6) becomes

$$\varepsilon_o k_{z1} + \varepsilon_1 k_{zo} = 0 \tag{10.13}$$

which leads for p polarised waves to the dispersion relation /10.3,3a/

$$k_x = \frac{\omega}{c}\sqrt{\frac{\varepsilon_o\varepsilon_1}{\varepsilon_o+\varepsilon_1}}\ , \tag{10.14}$$

see the curve $k_p d \gg 1$ in Fig. 10.3, valid for all k_x values.

If the values ε_o and ε_2 are different, asymmetric case, we obtain for sufficiently thick films ($i = 0$ or 2)

$$k_x = \frac{\omega}{c}\sqrt{\frac{\varepsilon_i\varepsilon_1}{\varepsilon_i+\varepsilon_1}}\ . \tag{10.15}$$

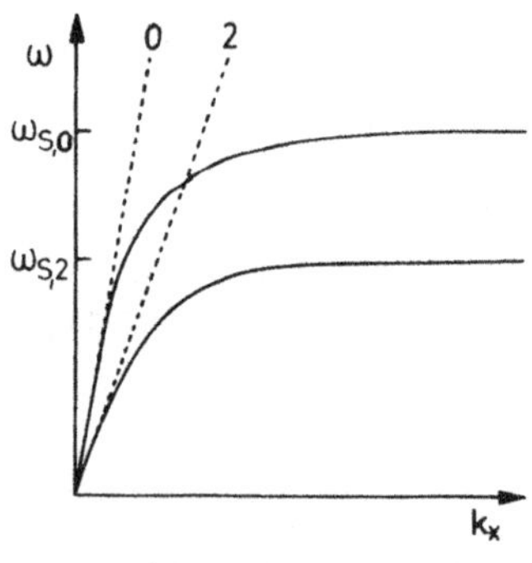

Fig. 10.4. Dispersion relations in the asymmetric layer system

As Fig. 10.4 shows the asymptotic values ω_{si} for $k_x \to \infty$ as well as the slopes for small k_x: $\sqrt{\varepsilon_i}\ \omega/k_x$ are different for the SP at the two boundaries.

Damping of the waves. As in general $\varepsilon_1(\omega)$ is complex $(\varepsilon_1' + i\varepsilon_1'')$, k_x has an imaginary part. For $\varepsilon_1'' \ll |\varepsilon_1'|$ and $\varepsilon_o = 1$ we obtain from (10.14) /10.4/

$$\mathrm{Im}(k_x) = \frac{\varepsilon_1''}{2\,\varepsilon_1'^2}\,\frac{\omega}{c}\left(\frac{\varepsilon_1'}{\varepsilon_1'+1}\right)^{3/2} \tag{10.16}$$

$$\mathrm{Re}(k_x) = \frac{\omega}{c}\left(\frac{\varepsilon_1'}{\varepsilon_1'+1}\right)^{1/2} \tag{10.17}$$

so that

$$\exp\{-\mathrm{Im}(k_x)\cdot x\} = \exp(-\frac{x}{\ell}) \tag{10.18}$$

determines the damping of the amplitude of the oscillations.

In the case of a free electron gas one obtains

$$\ell^{-1} = \frac{1}{2}\frac{1}{c\cdot\tau}\frac{\omega^2}{\omega_p^2} \qquad \text{for } \omega \ll \omega_p \tag{10.19}$$

which reaches on silver in the case of long wavelenghts (10 μm) values of about 1.5 cm, in the visible region about 100 μ.

Spatial extension of the field. The electric field extends into the z direction as $\exp(i\,k_{zi}\,z)$ with imaginary k_{zi} so that for *large* values of k_x compared to k_p, see (10.4), these fields decay in both media, in the plasma as well as in the dielectric, as

$$\sim \exp(-k_x|z|). \tag{10.20}$$

This means that they are strongly concentrated around the boundary. For *small* k_x values the fields decrease into the interior of the film as

$$\sim \exp\left(-\sqrt{|\varepsilon_1|}\,\frac{\omega}{c}\,z\right). \tag{10.21}$$

This means that the field disappears rapidly inside the plasma ($|\varepsilon_1|$ is large) and very slowly outside the film ($|\varepsilon_o| = 1$). In the case of rather long radio waves we call this the Sommerfeld wave.

Radiative modes. If k_{zo} is a real quantity, oscillatory solutions for $|z| > d/2$ are expected. This means that light is propagating away from the boundaries leading to a radiation damping in addition to the inner damping given by $\mathrm{Im}(\varepsilon_1)$. With real k_{zo} (10.2) is written, symmetric case assumed,

$$L^+ = \varepsilon_1 k_{zo} - \varepsilon_o k_{z1}\; i\;\mathrm{tg}(k_{z1}d/2) = 0 \tag{10.22}$$

$$L^- = \varepsilon_1 k_{zo} + \varepsilon_o k_{z1}\; i\;\mathrm{tg}(k_{z1}d/2) = 0 \tag{10.23}$$

for p polarised waves.

As the p polarised waves are strongly damped for large k_x values (see below), we discuss them only for small k_x.

For very small k_x ($k_x \sim 0$) (10.22) transforms into

$$\sqrt{\varepsilon_0}\ \mathrm{tg}\ \sqrt{\varepsilon_1}\ \frac{\omega}{c}\frac{d}{2} + i\ \sqrt{\varepsilon_1} = 0. \tag{10.24}$$

Assuming $\mathrm{Im}(\varepsilon_1) = 0$ (10.24) requires

$$\varepsilon_1' = 0 \tag{10.25}$$

which in case of a free electron gas is identical with $\omega = \omega_p$. This important mode discussed by FERRELL /10.5/ and marked $m_p = 0$ has been detected in several experiments as we shall see.

The dispersion relation can be derived from (10.22) /10.1c,6/ [up to terms quadratic in (k_x/k_p)] for $\frac{1}{2} k_p d \gg 1$ as

$$\frac{\omega}{\omega_p} = 1 + \frac{1}{2}\left(\frac{k_x}{k_p}\right)^2. \tag{10.26}$$

The damping under this condition is given by

$$\frac{1}{\omega_p\tau} = \frac{1}{k_p d}\left(\frac{k_x}{k_p}\right)^2. \tag{10.27}$$

For $\frac{1}{2} k_p d \ll 1$

$$\frac{\omega}{\omega_p} = 1 + \frac{1}{8}\,(k_p d)^2\left(\frac{k_x}{k_p}\right)^2 \tag{10.28}$$

and

$$\frac{1}{\omega_p\tau} = \frac{k_p d}{4}\left(\frac{k_x}{k_p}\right)^2, \tag{10.29}$$

see Fig. 10.5.

In both cases the eigenfrequency increases with k_x^2 in the region $k_x/k_p = \sin\theta$ up to ~ 0.7, see Fig. 10.5. The damping grows strongly with k_x, so that the observation of the dispersion of this excitation is difficult.

The mode ($m_p = 0$) shows no radiation damping for $k_x = 0$, since there exists only a field $E_z = \mathrm{const.}$ in the interior of the plasma slab, but not outside ($E_x = E_{z0} = 0$) as in a plane condensor of infinite dimensions. With increasing k_x the values of E_x and E_{z0} get finite and radiation damping starts, see Fig. 10.6.

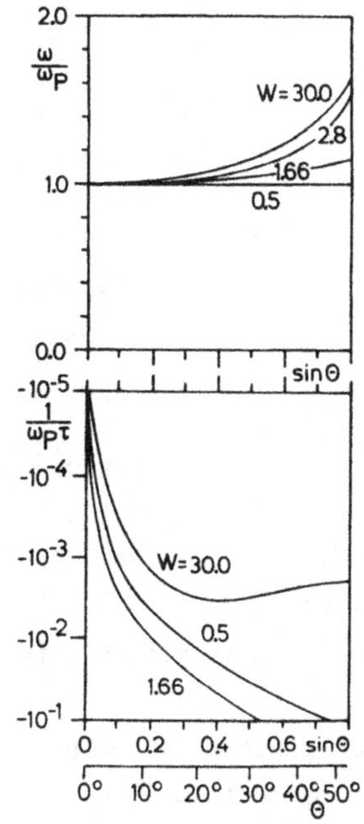

Fig. 10.5. The dispersion of the radiative modes and their damping as a function of $\sin\theta$ ($k_x = (\omega/c)\sin\theta$ with $\omega \sim \omega_p$). $W = k_p d$

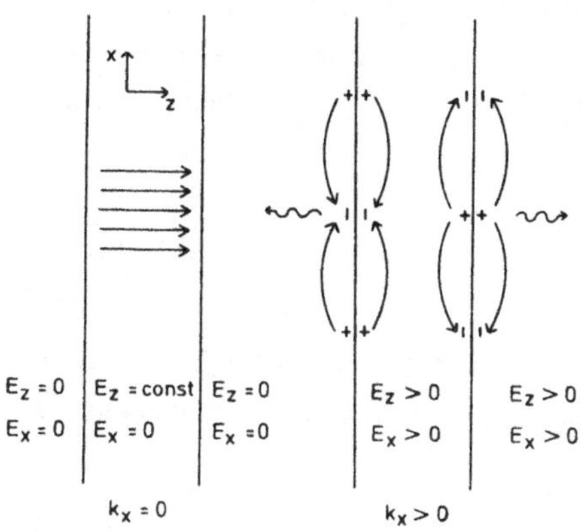

Fig. 10.6. Field configuration of the mode $m_p = 0$ for $k_x = 0$ and for $k_x > 0$

The dispersion relations (10.22 and 10.23) contain trigonometric functions so that solutions with higher harmonics of the $m_p = 0$ mode exist. For small ε_1, see (10.22), these solutions behave roughly as

$$\mathrm{tg}\,(k_{z1}d/2) = 0 \tag{10.30}$$

or

$$k_{z1}d \sim 2m\pi \tag{10.31}$$

and as

$$\mathrm{ctg}\,(k_{z1}d/2) = 0 \tag{10.32}$$

or

$$k_{z1}d \sim 2(m+1)\pi . \tag{10.33}$$

These field configurations represent standing waves in the z direction; m means the number of knots of E_z inside the film. Contrary to the tg ($m_p = 0$) mode of (10.22) the modes $m_p > 0$ are strongly damped even for $k_x = 0$. Consequently it is rather difficult to observe them.

Coupling of plasmons with photons. If k_{zi} is real or $k_x < \omega/c$ photons can be produced by plasmons. The conservation of momenta in this emission process is illustrated in Fig. 10.7: light with $k_\ell = \omega/c$ is emitted under the angle θ by a plasmon of momentum $\hbar k_x < \hbar\omega/c$; the lacking momentum $\hbar k_z$ is taken from the surface so that $(\hbar k_z)^2 + (\hbar k_x)^2 = (\hbar\omega/c)^2$ is fulfilled. This process is not possible for plasmons with $k_x > \omega/c$ or: nonradiative plasmons are not damped by radiation losses.

However if these nonradiative plasmons propagate on, e.g., sinusoidally corrugated surfaces they can exchange momenta $\pm\, n\,\frac{2\pi}{a}$ with the surface (n integer and a grating constant). The resulting wave vector

$$k_x' = k_x - n\,\frac{2\pi}{a} < \omega/c \tag{10.34}$$

allows the emission of radiation as displayed in Fig. 10.8 (process 1 → 2). In case of a rough surface we can apply similar consideration to every Fourier component of

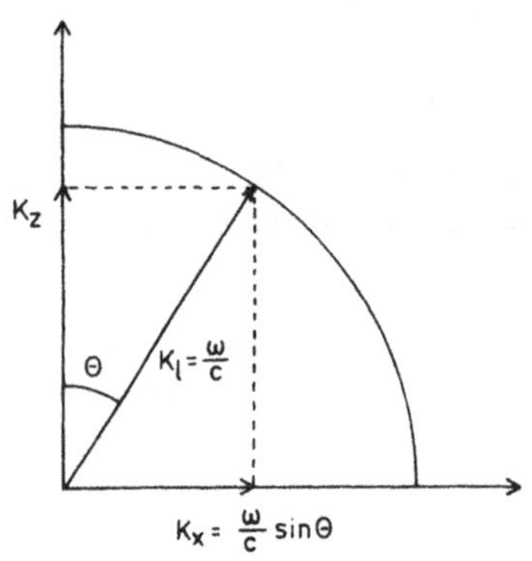

Fig. 10.7. The emission of a photon $\hbar\omega$ by a radiative plasmon with momentum $\hbar k_x$ and energy $\hbar\omega$

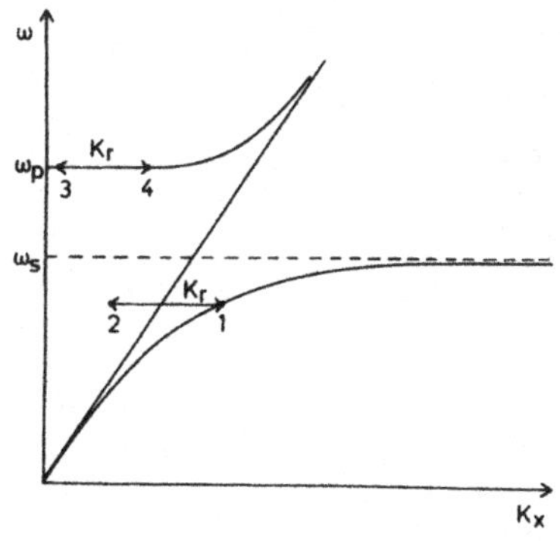

Fig. 10.8. Sketch of the coupling of radiative plasmons (process 3⇌4) and of nonradiative plasmons (process 1⇌2) with photons via surface roughness

the roughness spectrum. Thus nonradiative SP on surfaces of a perturbated profile radiate light /10.39/. The angular distribution of the emitted light can be used to obtain information of the roughness structure, see Sect. 10.4.

Naturally the reverse process can also take place and allows one to excite nonradiative surface plasmons, e.g., by irradiating a metallic grating with light, see Fig. 10.8 (process 2 → 1), "grating coupling". Besides the inner damping an additional radiation damping of these plasmons thus comes into play.

If we apply the same considerations to the radiative mode, we see that light of frequency ω_p directed on a rough film produces light of frequency ω_p radiating into all directions, since the dispersion relation is nearly a horizontal line and the vector $k_x' = k_x \pm k_r$ lies thus on the dispersion line too, see Fig. 10.8 (process 3 ⇌ 4). This phenomenon is called the plasma resonance emission.

Nonradiative and radiative light modes. The dispersion relations (10.22,23) can also be derived by a simple consideration: We assume a slab with a dielectric function $\varepsilon_1(\omega)$ covered on both sides with different media ε_0 and ε_2, see Fig. 10.2. The existence condition of an eigenmode is fulfilled, if light starting from point (1) arrives at point (2) with a phase difference $2\pi m$, m integer (Fig. 10.9c). With k_{z1}, the z component of the wave vector inside the slab and ϕ_0, ϕ_2 the phase shifts of the light being reflected at the boundaries 1/0 and 1/2, the condition of constructive interference is given by

$$2\, k_{z1} d = \phi_0 + \phi_2 + 2\pi m. \tag{10.35}$$

For ϕ_i (i = 0.2) follows from the Fresnel coefficients /10.1c/

$$\phi_i = 2 \text{ arctg } \frac{\varepsilon_1}{\varepsilon_i} \frac{k_{zi}}{k_{z1}} \; ; \qquad \phi_i = 2 \text{ arctg } \frac{k_{zi}}{k_{z1}} \; . \tag{10.36}$$

(p polarisation) (s polarisation)

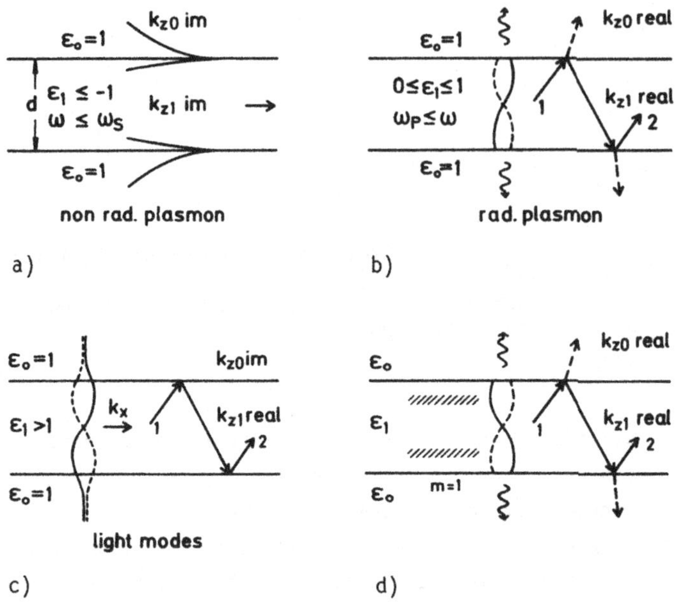

Fig. 10.9 a-d. Comparison of plasmons and guided light modes in thin films

We obtain the dispersion relation (10.2,3) for both types of plasmons by inserting ϕ_i of (10.36) into (10.35) either for p or s polarised modes: The equation for non-radiative plasmons if k_{z1} and k_{zi} are imaginary and the equation for radiative plasmons if k_{z1} and k_{zi} are real (i = 0.2).

Additional solutions of (10.35) [and therefore also of (10.2)] are guided light modes propagating in a dielectric film with real $\varepsilon_1 > 1$, e.g., in films of quartz, glass, etc. In the case of real k_{z1} and imaginary k_{zi} we have the nonradiative light modes, see Fig. 10.9c: The electromagnetic wave in the z direction is a standing wave with small amplitudes at the boundary and m knots in the interior. In contrast to the nonradiative plasmons which have their maximum amplitude at the boundary, see Fig. 10.9a, the guided light modes have their maximum or maxima in the interior of the slab. This field distribution travels with the phase velocity ω/k_x in the x direction.

In Fig. 10.10 the d(θ) dependence is drawn for a symmetric system ($\varepsilon_0 = \varepsilon_2$) and for a given frequency ω. For every thickness d a number of light modes exist for different values of $\sin\theta = k_x c/\omega$, different for p and s modes since the values of the ϕ_1 in (10.35 and 10.36) depend on the polarisation. The usual plot $\omega(k_x)$ of the dispersion relation of light modes is given in Fig. 9.3; here the modes lie between the light lines $\omega/k_x = c$ ($\varepsilon_0 = 1$) and $c/\sqrt{\varepsilon_1}$.

If in the asymmetric case one of the adjacent media, e.g., ε_2, is a metal we have a layer system: metal (ε_2) / dielectric (ε_1) / dielectric (ε_0). Then in addi-

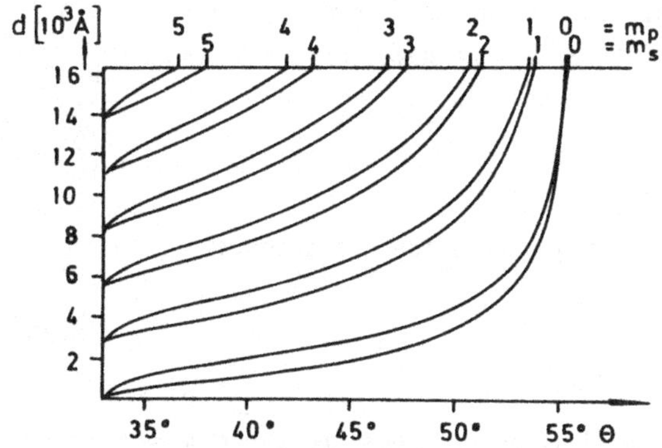

Fig. 10.10. The relation between film thickness d and the angle θ for s and p polarised light modes. The curves are calculated for λ = 6328 Å, n (film) = 1.52 and n (prisma) = 1.839 /10.8/

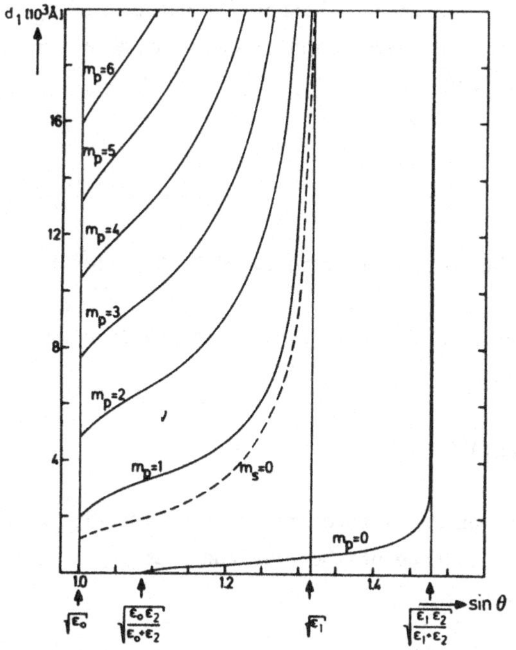

Fig. 10.11. The relation between the film thickness d_1 and the angle θ for different polarisation of the light modes for ε_0 = 1 (air) and ε_1 a transparent dielectric material deposited on a metal substrate $\varepsilon_2(\omega)$. m_p = 0 is the SP mode; the curves for s and p polarised light modes start now at different thicknesses as indicated for m_s = 0. (10.37)

tion to the light modes in (ε_1) we expect the SP in the interface $\varepsilon_2/\varepsilon_1$. The dispersion relation is drawn in Fig. 10.11: If $d_1 = 0$ we have for the SP mode ($m_p = 0$)

$$\sin\theta = \frac{k_x}{\omega/c} = \sqrt{\frac{\varepsilon_o \cdot \varepsilon_2}{\varepsilon_o + \varepsilon_2}} \tag{10.37}$$

as is evident. With increasing d_1 the dispersion relation converges to

$$\sin\theta = \frac{k_x}{\omega/c} = \sqrt{\frac{\varepsilon_1 \cdot \varepsilon_2}{\varepsilon_1 + \varepsilon_2}} \tag{10.38}$$

as is derived above. In between these two limiting θ values the curve $m_p = 0$ in Fig. 10.11 describes the thickness dispersion, i.e., the dependence of θ (or k_x) on the coating thickness d for fixed ω.

We obtain the "radiative" guided light modes of the slab for real k_{zo} or (and) k_{z2}, see Fig. 10.9d. As the condition for total reflection is not fulfilled, a certain radiation loss takes place as with the radiative plasmons, see Fig. 10.9b, so that, e.g., the transport of information over large distances is not possible with these modes in contrast to the nonradiative guided light modes. For more details on these guided light modes see /10.9,10/.

10.2 Excitation of Surface Plasmons

As an electromagnetic phenomenon the described modes can be excited by the fields of charged particles and photons.

Excitation by electrons. Electrons entering a crystal transfer energy $\Delta E = \hbar\omega$ and momentum $\hbar q$ to the excitation with

$$(\hbar q)^2 = (\hbar k_{el}\theta)^2 + (\hbar \frac{\omega}{v})^2 = (\hbar q_\perp)^2 + (\hbar q_\parallel)^2 \tag{10.39}$$

with k_{el}: wave vector of the incoming electron, θ: scattering angle of the electron, $\hbar\omega$: transferred energy. In case of normal incidence of the electron beam the component of k_{el} parallel to the surface is given by

$$k_x = q_\perp = k_{el} \cdot \theta \; ; \tag{10.40}$$

varying θ allows one to scan over the dispersion relation as Fig. 10.12 demonstrates. Radiative as well as nonradiative plasmons can thus be excited. The excitation is detected by measuring the energy losses of the electrons in an energy analyser as described in Chap. 11.

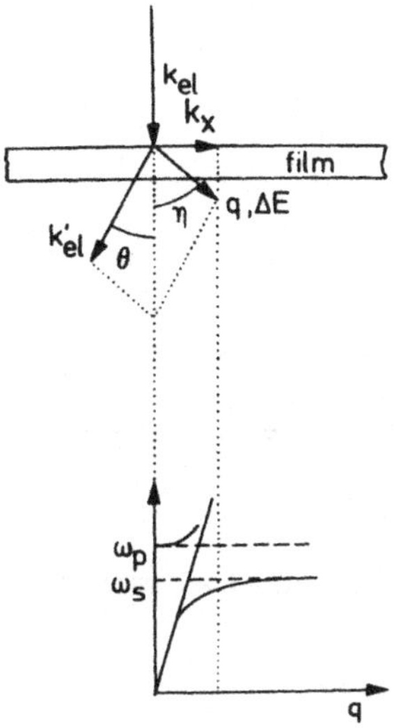

Fig. 10.12. Sketch of the excitation of radiative and nonradiative plasmon with wave vector k_x by an electron beam. $\hbar q$ is the momentum transferred from the electron to the system

Excitation by light. Light directed onto a surface can excite radiative SP's directly as discussed in Fig. 10.7. This is not possible with nonradiative SP's since one cannot match frequency and wave vector of the surface plasmon. Two methods allow one to overcome this difficulty:

(a) The grating coupler, see Fig. 10.8. Here a corrugated surface, e.g., a grating with periodicity a, is irradiated with light of frequency ω and wave vector $k_\ell = \omega/c$. The wave vector in the boundary k_x can be changed into $k'_x = k_x \pm \frac{2\pi n}{a}$, n integer. Thus the dispersion relation (k'_x) can be fulfilled, see Fig. 10.8. The excitation is detected by a deep minimum of the reflected (or transmitted) light since the plasmons absorb energy.

(b) Prism method. Light falls from the dielectric material ε_0, e.g., quartz, on the boundary of a plasma slab $\varepsilon_0/\varepsilon_1$, see Fig. 10. 13; the wave vector component parallel to the surface is given by $k'_x = \sqrt{\varepsilon_0}\,(\omega/c)\ \sin\theta_0$. If $\sqrt{\varepsilon_0}\ \sin\theta_0 \geq 1$, the vector k'_x lies right from the light line and total reflection at the boundary 0/1 takes place so that an inhomogeneous wave with the phase velocity $c/(\sqrt{\varepsilon_0}\ \sin\theta_0) < c$ runs along the boundary $\varepsilon_0/\varepsilon_1$. This inhomogeneous wave can excite nonradiative SP's at the boundary $\varepsilon_1/\varepsilon_2$ if the thickness of medium ε_1 is not too large. ε_1 can be the plasma and ε_2 a dielectric, e.g., air, or vice versa. In the experiment medium (ε_0) is given by form of a prism or a half cylinder. Entering the medium (ε_0) the light gains the additional momentum $\Delta k_x = (\sqrt{\varepsilon_0}-1)\,\frac{\omega}{c}\sin\theta_0$ so that it can excite the nonradiative plasmons. The resonance case is detected by a deep minimum in the totally reflected beam since the plasmon absorbs energy, as in the case of grating coupling. Therefore the prism method is often called the ATR method (attenuated total reflec-

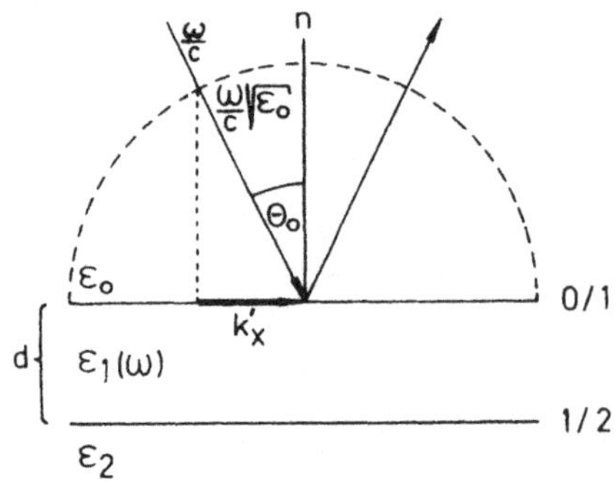

Fig. 10.13. Sketch of the prism method

tion). Then it would be consequent to call the grating coupling method AR method (attenuated reflection, see Fig. 10.29).

10.3 Results on Radiative Plasmons

Excitation by electrons, plasma radiation. As Fig. 10.12 demonstrates electrons can excite radiative plasmons of frequency ω_p if $\hbar k_x \lesssim \hbar k_p$ is transferred to the surface. These plasmons decay into light of the same frequency ω_p. This "plasma radiation" has been predicted by FERRELL /10.5/.

If an electron passes the boundary of two media with different dielectric functions an electromagnetic pulse is produced; this "transition radiation" has a broad continuous spectrum. If one of these media is a metal with a dielectric function which goes through zero, a pronounced intensity peak at $\omega = \omega_p$ is superposed, if damping is small. This light is called "plasma radiation". The nonrelativistic intensity of the plasma radiation or the number of photons of energy $\hbar\omega$ per electron and energy interval $d(\hbar\omega)$ emitted into the solid angle $d\Omega$ of the direction θ in a thin film of thickness d can be approximated for small ε'', $\sin^2\theta \ll \varepsilon''$ and $k_p d \ll 1$ by, see /10.11/

$$\frac{\partial^3 N}{\partial(\hbar\omega)\partial^2\Omega} = \text{const.}\ \frac{\gamma_d^2}{4(\omega-\omega_p)^2 + \gamma_d^2} \ . \tag{10.41}$$

The damping coefficient γ_d is the sum of the inner damping

$$\gamma_i/\omega_p = \frac{1}{\omega_p\tau_i} = \varepsilon''(\omega_p) \tag{10.42}$$

and the radiation damping

$$\gamma_r/\omega_p = 1/\omega_p\tau_r = \frac{1}{2}\,k_p d\,\sin\Theta \cdot \mathrm{tg}\Theta. \tag{10.43}$$

Equation (10.43) shows that the radiation damping is increasing with the thickness d of the irradiated film and the angle θ of the emitted light against the film normal. For comparison with the experimental results the relativistic formula is used /10.12-14/.

Experiments on thin silver films of about 500 Å thickness by STEINMANN /10.15/ and BROWN et al. /10.16/ have verified that irradiation of a film with electrons of 50-100 keV produces plasma radiation with a peak at about 3.8 eV (3250 Å) in agreement with the calculated value. The intensity of the radiation on a thin film is higher than on a boundary of a semiinfinite metal. Figure 10.14 displays recent measurements of the spectral distribution of the absolute light intensity at $\theta = 20^{o}$ of a 600 Å silver film irradiated with 80 keV electrons (dashed line). The open and full circles are calculated with the relativistic formula using a dielectric function measured at the same film /10.17/. These experiments have been extended to metals with a more free electron gas character like Al, Mg, Be, etc. Measured values of the plasma energy ΔE_p and the halfwidth $(\Delta E_p)_{1/2}$ of the plasma radiation peak are listed in Table 10.1 which are in good agreement with those obtained with other methods, see Table 7.1.

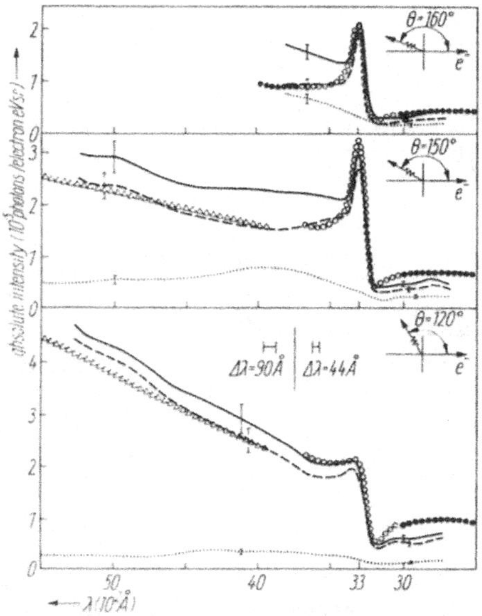

Fig. 10.14. Spectral distribution of the (absolute) intensity of the transition radiation of a silver film of 600 Å thickness bombarded with 80 keV electrons. The plasma radiation peak at ω_p surpasses the background of the transition radiation. The full line is the measured p polarised, the dotted line the measured s polarised intensity; the dashed line represents p minus s polarised intensity, which is compared with the calculated values (circles) /10.17/

Table 10.1. ΔE_p and halfwidth $(\Delta E_p)_{1/2}$ of the plasma radiation peak

Material	ΔE_p [eV]	$(\Delta E_p)_{1/2}$ [eV]	Ref.
Ag	3.78		/10.18/
Al	14.9		/10.19/
Be	18.1 ± 0.4	5	/10.20/
Cd	9.1	0.9	/10.21/
In	11.8		/10.21/
Mg	10.4	0.6	/10.22/

The dispersion of the plasmon peak measured by the displacement of the plasma frequency with the angle θ, (10.26), as well as the increase of the halfwidth of the peak due to growing radiation damping has been confirmed by the experiment /10.20/. Extrapolation of the halfwidth $(\Delta E_p)_{1/2}$ to $\theta = 0$ gives the inner damping of the plasmon $(1/\omega_p\tau_i)$. For Mg a comparison of its value with those obtained by other methods is given in Table 10.2 which shows the consistency of the data.

Table 10.2. ΔE_p and halfwidth $(\Delta E_p)_{1/2}$ of the plasma radiation of Mg measured different methods

ΔE_p [eV]	$(\Delta E_p)_{1/2}$ [eV]	Ref.
10.4	0.58 [plasma radiation (20 °C)]	/10.22/
10.2	0.6 [plasma resonance, photoeffect (20 °C)]	/10.23/
10.2	0.75 [plasma resonance absorption (-196 °C)]	/10.24,25/
10.5 ± 0.1	0.6 [light reflection experiments]	/10.26/

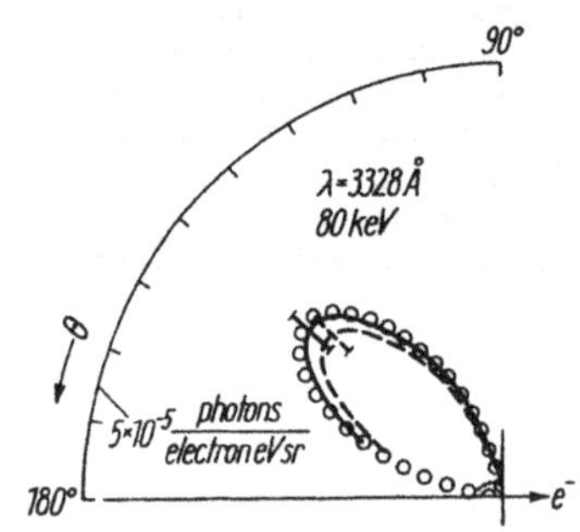

Fig. 10.15. Angular dependence of the intensity of the silver plasma radiation. Solid line: measured p polarised intensity, dashed line: p minus s polarised intensity, circles: calculated values. The same distribution is observed in the other three quadrants /10.17/

The angular distribution of the radiation is reproduced in Fig. 10.15; it equals that of the transition radiation ($I \sim \sin^2\theta \cos^2\theta$), (Hertz dipolelike).

The polarisation of the radiation at normal incidence of the incoming electrons shall be in the plane of observation. The observation however gives a rather strong s polarised contribution, see Fig. 10.14. Reduction of the measured p intensity by the intensity of the s components, (p-s) in Fig. 10.14 maintains the good agreement of calculated and measured data. There is evidence that the s polarised contribution is partly due to a certain roughness of the irradiated surface /10.17,27/ and not only to the expected bremsstrahlung.

Detection of energy losses due to excitation of radiative plasmons. Electrons which have excited volume plasmons and radiative SP's cannot be discerned at small wave vectors because both excitations have the same energy. For larger wave vectors however they behave in a different way: the loss energy of the SP's increases with k_x^2, (10.26), and approaches the light line whereas the energy of the volume plasmons is approximately independent of k_x in this wave vector region. Using (1.19) the relative energy change between $k = 0$ and k_p amounts only to some 10^{-5} for volume plasmons, which is too small to be observed whereas the energy of the SP's increases by some eV for the 15 eV loss in the region $k \sim 0$ and k_p. To measure the dispersion of SP's one needs electron energy loss arrangements with much higher angular resolution than those usually used. By electron optical means beam apertures of less than 10^{-5} rad for 75 keV electrons have been realized /10.28/ which correspond to $k_x < 3 \cdot 10^{-3}$ Å^{-1}. This value is small compared to $k_p = 2\pi/\lambda_p = 7.5\ 10^{-3}$ Å^{-1} for Al, so that the dispersion relation of radiative SP's could be measured up to $\sim k_p$ in agreement with the calculation.

Excitation by light. Radiative plasmons can be excited by light directed onto a smooth plasma film. Scanning the frequency of the incident light a resonance at ω_p should be observed. The momentum conservation is fulfilled since $k_x \leq \omega_p/c$, see Fig. 10.7. This phenomenon has been observed as (a) plasma resonance absorption, (b) plasma resonance emission.

a) Plasma resonance absorption. Figure 10.16 is a sketch of the experiment. Light falls under an angle θ on the plasma film. The transmitted light intensity is measured as function of its frequency. By using s polarised light - the electric vector oscillates perpendicular to the plane of incidence - the transmittance $T = I_T/I_0$ of a 300 Å silver film is observed as shown in Fig. 10.17a /10.29/. With p polarised light (electric vector in the plane of incidence) however a minimum of T is detected if the angle of incidence (θ) is different from zero, Fig. 10.17b. The energy at which the minimum is observed equals approximately 3.8 eV, the volume plasmon energy of silver. These observations at Ag /10.29-32/ have been verified with different materials like K /10.33/, Mg /10.34/, and Al /10.35/ which behave more like a free electron gas. The normal component of the electric field E_n of the incident light

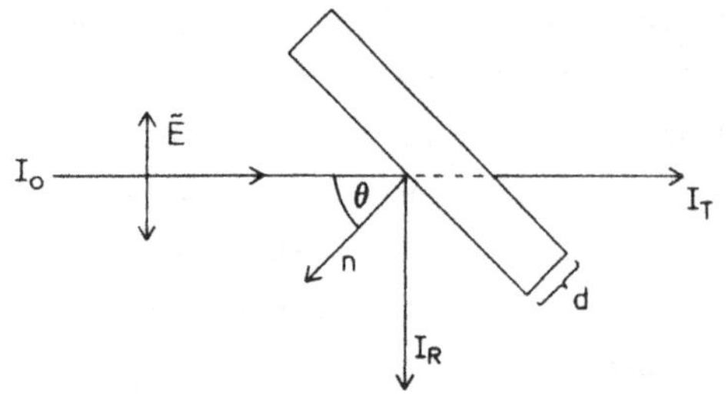

Fig. 10.16. Schematic figure of the experimental arrangement to observe plasma resonance absorption and emission

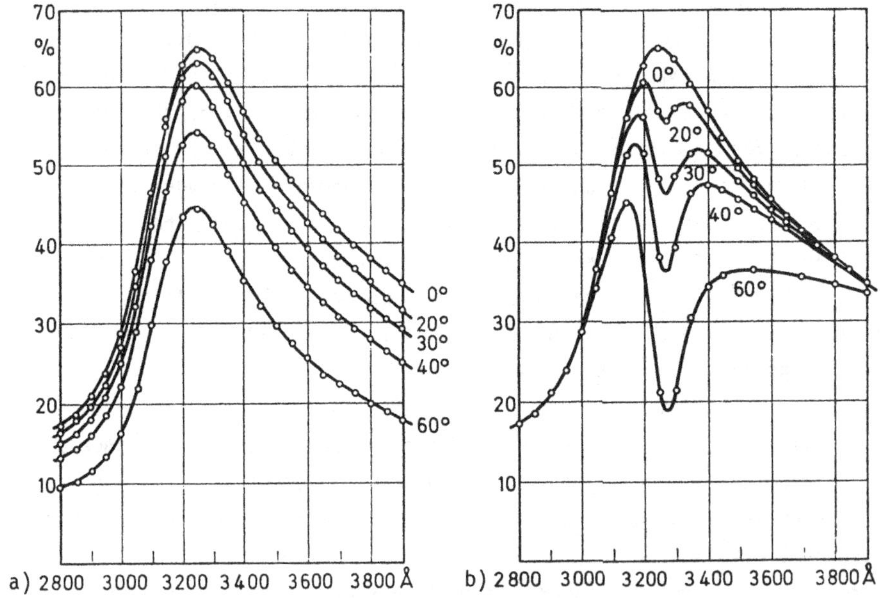

Fig. 10.17a,b. Transmission of s polarised (T_s) (left) and p polarised light (T_p) (right) of a silver film 300 Å thick for angles of incidence from 0 to 60° /10.29/

excites surface plasma oscillations and thus reduces the transmittance. Good agreement has been found between experimental and theoretical transmittance curves.

The plasma oscillation can transfer its energy to single electrons which leave the film as "photoelectrons" if the work function of the material is smaller than ΔE_p. The conservation of momentum is regulated by the surface. Experiments on thin Al films demonstrated a strong emission of electrons if the frequency of the incident light goes through $\Delta E_p/\hbar$ /10.35/.

b) Plasma resonance emission. If in an arrangement as sketched in Fig. 10.16 the reflected intensity is measured for an angle different from θ (angle of incidence) a peak at ΔE_p is observed in this nonspecularly reflected light when scanning the wavelength /10.36-38/, see Fig. 10.18. The angular intensity distribution is that of a Hertz dipole with its axis parallel to the film normal. The emitted light is mainly polarised in the plane of observation.

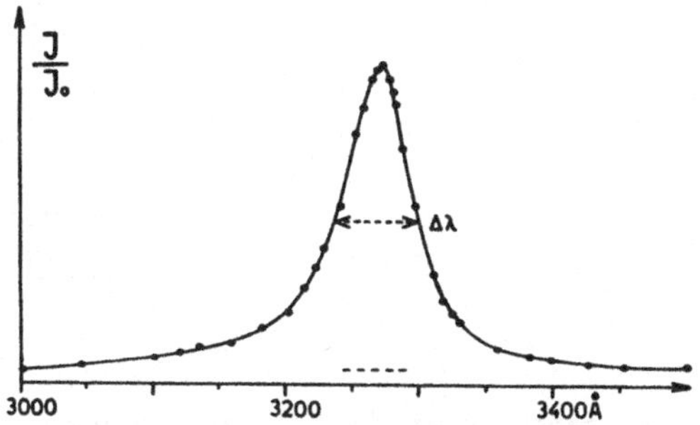

Fig. 10.18. Plasma resonance emission from a silver film (500 Å thick). Relative light intensity I/I_0, emitted at $\theta = 40^0$ is plotted as function of the wavelength of the incoming light (I_0) incident at an angle of 23^0. The halfwidth of ~ 60 Å corresponds to about 70 meV /10.37/

The interpretation of this observation is given by Fig. 10.8: the roughness of the film represented by k_r couples the different directions of emission $k_x' = (\omega/c)\cdot \sin\theta' = k_x + k_r$ so that light of frequency ω_p is scattered into all directions as long as the dispersion relation is fulfilled /10.39,40/. (For frequencies different from ω_p scattered light is observed too but its intensity is several orders smaller than in the resonance case.)

If the light is scattered from the interior of the metal due to spatial variations of $\varepsilon(\omega)$, its intensity is proportional $1/\varepsilon(\omega)$ and a maximum at $\omega = \omega_p$ ($\varepsilon \sim 0$) is expected. Scanning the frequency of the incoming light, a resonance curve like that in Fig. 10.18 is measured. This result and further experiments /10.40/ support the interpretation of the resonance emission by inner roughness.

In contrast the light emitted by nonradiative plasmons propagating along a rough surface is produced by surface roughness, see Sect. 10.4.

10.4 Results on Nonradiative Plasmons

Excitation by electrons. As Fig. 10.12 demonstrates electrons can excite nonradiative SP in the region from very low values of k_x up to large k_x. The energy losses ΔE are detected by a suitable energy analyser as described in Chap. 11.

Probability of surface excitation. If an electron passes the surface of a plasma film of thickness d at normal incidence the differential probability of losing the energy ΔE and being scattered by an angle θ into the solid angle $d\Omega$ has been calculated by RITCHIE /10.2/ neglecting retardation ($k_x > \omega/c$) to

$$\frac{\partial^3 W_s}{\partial\Delta E \partial^2\Omega} = \left(\frac{e}{\pi\hbar v}\right)^2 \cdot k_{el}^{\;2} \cdot \mathrm{Im}\left[\frac{(\varepsilon-\varepsilon_0)^2}{\varepsilon\varepsilon_0} \; \frac{2k_x^{\;2}}{q^4} \cdot R\right] \tag{10.44}$$

$$R = \frac{\sin^2 \frac{\omega}{v}\frac{d}{2}}{iL^+} + \frac{\cos^2 \frac{\omega}{v}\frac{d}{2}}{iL^-} \; . \tag{10.45}$$

ε_0 means the dielectric function of the medium neighbouring the plasma film and v the electron velocity; $L^{\pm}$ are identical with (10.6), neglecting retardation. Coupling of the oscillations at the two boundaries in general can be neglected if in an experiment the film thickness is so large that

$$\frac{1}{2}\, k_x d = \frac{1}{2}\, k_{el}\theta d \gg 1 \tag{10.46}$$

or

$$\theta d \gg 10^{-2} \tag{10.47}$$

with 50 keV electrons. This allows one to replace R by

$$R = \frac{1}{k_x}\,\frac{1}{\varepsilon+\varepsilon_0} \; . \tag{10.48}$$

For large k_x or $\varepsilon \sim -\varepsilon_0$, see (10.8), we obtain from (10.44) the expression

$$\frac{\partial^3 W_s}{\partial\Delta E \partial^2\Omega} = \mathrm{const} \cdot \mathrm{Im}\left[\frac{(\varepsilon-\varepsilon_0)^2}{\varepsilon\varepsilon_0(\varepsilon+\varepsilon_0)}\right] \cong \mathrm{const} \cdot 4\; \mathrm{Im}\left[\frac{-1}{\varepsilon+\varepsilon_0}\right] \tag{10.49}$$

which has its maximum at $\varepsilon = -\varepsilon_0$, the frequency condition of the surface plasmons, see Fig. 10.19.

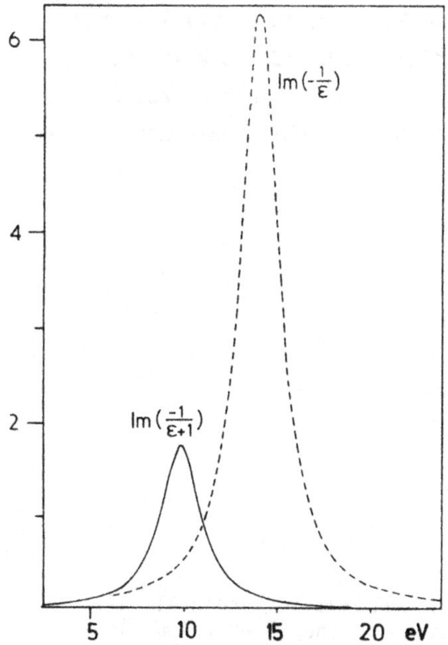

Fig. 10.19. The surface loss function Im(-1/1+ε) and the volume loss function Im(-1/ε) ($\hbar\omega_p$ = 15 eV, $\hbar/\tau$ = 2,5 eV, free electron gas)

Equation (10.49) can be splitted into $(\varepsilon-3\varepsilon_o)/\varepsilon_o \cdot 1/(\varepsilon+\varepsilon_o)$, the surface term, and $+1/\varepsilon$ which we have neglected above. The latter contributes to the volume term with an opposite sign. Its influence depends on film thickness and on the angular width of the beam, see the discussion in /10.42/. With the dielectric function of a free electron gas, $\omega_s = \omega_p/\sqrt{1+\varepsilon_o}$ is obtained, see (10.9b). Interband transitions displace this value. If $\varepsilon_o'' > 0$ damping of the surface loss takes place in addition to that caused by ε''.

In the region of small values of k_x ($k_x < \frac{\omega}{c}$) retardation cannot be neglected in the calculation of the loss probability. The relativistic relation of KRÖGER /10.14/ describes the appearance of transverse excitations as the Cerenkov radiation and of guided light modes, see Chap. 9. In addition it allows one to calculate more exactly the intensity of the surface contributions in the low energy region of the loss spectrum at small k_x values, see, e.g., /10.54/. The theoretical treatment has been extended to anisotropic crystals too, see /10.41/, and applied to a thin uniaxial slab (optical axis parallel to the surface normal).

Experimental values of surface losses. Figure 10.20 shows an energy loss spectrum of a thin Al film (200 Å, θ = 0.5 mrad); it displays besides the volume loss at 15 eV two surface losses $\hbar\omega^{\pm}$ demonstrating the splitting discussed above. The k_x dependence of ω^+ and ω^- has been measured with ELS at slightly oxidized Al films /10.43/.

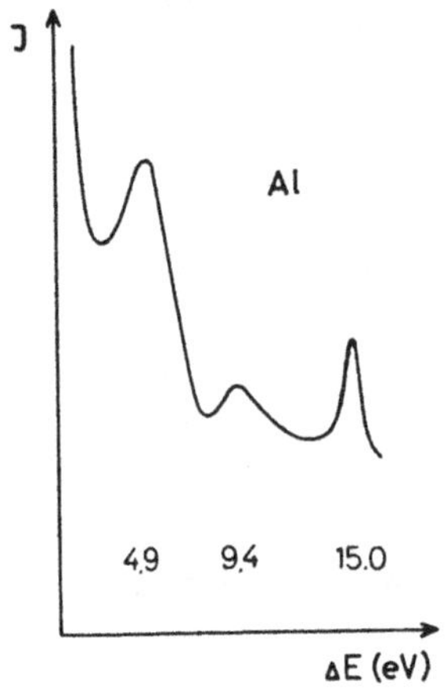

Fig. 10.20. Energy spectrum of a thin Al film (200 Å) which consists of the 15 eV volume loss and the surface losses 4.9 and 9.4 eV. The last two represent the 7 eV loss, split up into the doublet by coupling of the two surface waves. Observation at θ = 0 with an angular aperture of $1.6 \cdot 10^{-4}$ rad /10.42/

In general however thicker Al films of $\sim$ 500 Å thickness have been used. For these films the SP's at the two boundaries are decoupled and one surface loss has been observed. In Table 10.3 observed values of ΔE_s are compared with calculated $\hbar\omega_p/\sqrt{2}$ data. One finds a rather good agreement between calculated and observed data.

Since the surface loss energy depends strongly on k_x in the retardation region one needs a specification of k_x for the data in Table 10.3. In the experiments usually an aperture of $\leq 10^{-4}$ rad is applied, which is approximately twice the value of θ_p ($\theta_p = k_p/k_{el}$; for electrons of 50 keV and λ_p = 850 Å of Al: $\theta_p = 6 \cdot 10^{-5}$ rad). Measuring at θ = 0 with the above-mentioned aperture the value of ΔE_s is averaged over the frequency range of the retarded region. This value is close to the asymptotic one since the integration over θ accentuates the higher ω values.

Differences of the observed values ΔE_s and $\hbar\omega_p/\sqrt{2}$ are due to deviations of the dielectric function from that of a free electron gas. This is very pronounced for Ag and Au, e.g., the condition ε = -1 leads in Ag to 3.63 eV (ε = 0 : 3.78 eV) and in Au to 2.5 eV (ε = 0 : 6 eV) /10.46/.

The halfwidth of the SP can be determined in transmission experiments with less accuracy than that of the volume plasmons. In general the low intensity of the surface peaks makes it more difficult to separate them from the background and from the neighboured volume peak especially if measurements at larger q values are made.

Table 10.3. Some surface and volume plasmon energies ($\hbar\omega_p$ calculated values)

	Al	Mg	Li	Na	K	In[a]	Si	Ag
ΔE_s [eV]	10.6	7.15	3.95	3.85	2.60	8.7	10.8 ± 0.2	3.6 ± 0.02
$\Delta E_{s\ 1/2}$ [eV]	1.5-2	1.4-2	2-3	0.9-1.2	0.7-1	1.2	8.8	0.25
$\hbar\omega_p/\sqrt{2}$ [eV]	10.58	7.34	5.01	3.96	2.62	8.9	11.7	6.3
$\hbar\omega_p$ [eV]	15.7	10.9	8.02	5.95	4.29	12.5	16.6	9
Ref.	/10.43/	/10.43/	/10.43/	/10.43/	/10.43/	/10.44,45/	/10.45/	/10.60,60a/

[a] in solid state. These data are in very good agreement with those calculated from optical measurements /10.45,165/

More detailed experiments are desirable with films so thin that the intensity of the surface loss becomes comparable with (or larger than) that of the volume loss, see for example /10.51/. The result would be of interest for a more precise determination of the width of the surface losses and for questions as for example the dependence of their width on q, see Sect. 6.1, in analogy to the volume plasmons.

Up to now the comparison of the width of the SP loss peaks of Al and K /10.43a/ and Ag /10.54/ with calculated values using the dielectric theory demonstrated a rather good agreement within 10-20 %. Experiments in reflection are more accurate /10.45/ due to the enhancement of the SP peak by a factor $1/\cos\alpha$ (10.54). The measured width of the SP of In and Si, see Table 10.3, are also in good agreement with those calculated from *optical* measurements. This means that in good approximation the damping of the surface oscillations is described by the imaginary part of the dielectric function and no specific influence of the surface is observed.

Angular dependence of the surface loss intensity. As (10.44) shows, the differential loss probability has an angular dependence as k_x/q^4 or

$$\theta/(\theta^2 + \theta_{\Delta E}^2)^2 \tag{10.50}$$

which gives an intensity increasing with θ, passing a maximum at

$$\theta_{max} = \theta_{\Delta E}/\sqrt{3} \tag{10.51}$$

with $\theta_{\Delta E} = \Delta E/2E_0$ and decreasing for $\theta > \theta_{\Delta E}$ with

$$\frac{\partial^3 W_s}{\partial\Delta E\,\partial^2\Omega}\,(\theta > \theta_{max}) = \text{const.}\ \theta^{-3}. \tag{10.52}$$

The θ^{-3} dependency has been verified by several measurements on Mg /10.47,48/, Al /10.48/, and Ag, /10.48-50/. The values have been measured at films vaporized in a vacuum of a few 10^{-9} Torr on thin carbon films surrounded by a shield cooled with liquid N_2.

If one measures with apertures of several 10^{-4} rad, the integrated excitation is obtained practically. One calculates for a free electron gas with (10.44)

$$W_s = \frac{1}{1+\varepsilon_0}\,\frac{\pi e^2}{\hbar}\cdot\frac{1}{v} \tag{10.53}$$

for an electron passing one boundary of a plasma. Twice the value has to be taken if the electrons have to pass the two boundaries of a film. For 50 keV electrons W_s has a value of some 10^{-2}, which has been verified by the experiments /10.50/. This probability can be increased by nonnormal incidence of the electrons, so that (10.53) changes into, see (10.58),

$$W_s(\alpha) = W_s(0)/\cos\alpha. \tag{10.54}$$

If the value of W_s, the probability of producing *one* surface loss $\hbar\omega_s$, grows the probability of multiple losses $n\cdot\hbar\omega_s$ also increases and has to be taken into account. As we have seen in Sect. 4.3 a loss of $n\cdot\hbar\omega_s$ is excited with a probability

$$P_n = \frac{1}{n!} W_s^{\,n}\, e^{-W_s}. \tag{10.55}$$

This relation has been verified for SP's, see /10.45/.

The effect of the cosα factor is due to the strong interaction of the electrons hitting a surface at grazing incidence compared with that of an electron at normal incidence, where the loss probability is relatively small and multiples are not observed in general. In the case of multiple volume losses $n\hbar\omega_p$ due to the weak interaction (3.3) the primary electron suffers repeated single processes, spatially separated. This is not possible with SP except in thin films; here the electron may suffer 2 $\hbar\omega_s$ passing the two boundaries.

The strong and short interaction of the electron with the surface at small cosα, acting as a bang, contains besides the fundamental frequency ω_s a number $n\cdot\omega_s$ so that the levels $n\cdot\hbar\omega_s$ are occupied corresponding to the Poisson distribution /10.46a/ in agreement with the observation.

This sort of interaction of the electrons is different from that of light which acts in general as a plane wave with a definite frequency and wave vector on the solid.

The surface loss at low k_x values. Figure 10.3 shows that the frequency of the surface plasmons goes to zero with decreasing k_x. As has just been discussed it is necessary to work with very small apertures to measure this decrease. Experiments with a beam divergence of $0.5\cdot10^{-4}$ rad or less demonstrated clearly that the frequency decreases /10.43/, see also Figs. 10.22 and 10.24.

With special electron optics an angular aperture of less than 10^{-5} rad has been achieved, so that the dispersion curve could be measured down to rather low $\hbar\omega$ values, see Fig. 10.21 /10.28/. This influence of the retardation is easier to observe by inclining the film normal against the electron beam /10.52,53/. The retardation effect also has an important influence on the loss intensity for low k_x ($\leq k_p$) values, so that (10.44) cannot be used in this region as has been demonstrated for the silver loss spectrum /10.54/.

The SP at large k_x values. Whereas the dependence of $\omega(k_x)$ is found in agreement with the dielectric theory for $k_x \lesseqgtr k_p$, the theoretical situation for large k_x is still open.

The dependence $\omega(k_x)$ for $k_x \gg k_p$ is determined by the decrease of the electron density n(z) at the boundary of the metal. A realistic model of n(z) at the metal

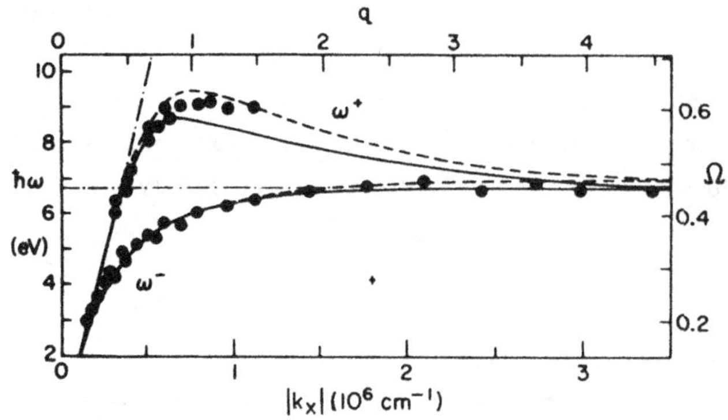

Fig. 10.21. Observed splitting of the Al surface loss (6.7 eV) into the ω^+ and ω^- mode as function of k_x /10.28/

boundary replaces the step function by a continuously decreasing function as has been shown for the static case (nonoscillating density profile) /10.55/. For $k_x \gg k_p$ the usual dispersion relation will therefore no longer be valid; this means that the relation ω_s = const. (ω_s asymptotic value for large k_x) will not hold if the decay of the plasmon field normal to the boundary $\exp(-|k_z|z)$ becomes comparable with the decrease of $n(z)$ which can be assumed as $\exp(-z/a)$. This condition gives $1/a \sim |k_z|$. $|k_z|$ can be replaced by k_x for large k_x, see (10.4), so that

$$k_x \sim 1/a \qquad (10.56)$$

will be a value for which changes of the usual dispersion relation should be essential. Assuming $a \sim 2$ Å one obtains $k_x \sim 0.5$ Å^{-1}, but deviations from ω_s = const. are expected already for smaller k_x values.

Experiments have been performed on the 10 eV Al loss by using fast (50 keV) and slow (100 eV) electrons. A certain difficulty with fast electrons is the strong decrease of the loss intensity with θ^{-3}. Up to θ = 2-2.5 mrad no appreciable change of $\hbar\omega_s$ has been observed, but a tendency of an increase of the energy value is seen /10.56/: triangles in Fig. 10.22. If thin Al films are used to reduce the influence of the volume peak, so that the positions of the surface loss can be better fixed for small peak heights, an increase of the loss energy to approximately 12 eV for q about 0.5 Å^{-1} has been observed /10.51/, open circles in Fig. 10.22. This is in rough agreement with the above estimation.

Results of reflection experiments with slow electrons on monocrystals of Al in different crystal directions /10.57/ which cannot be evaluated as straightforwardly as the experiments with fast electrons also demonstrated an increase of $\hbar\omega_s$ which agrees qualitatively with the results of ELS with fast electrons.

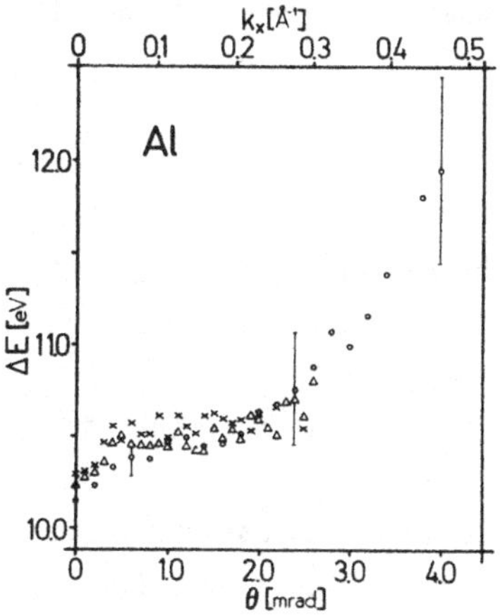

Fig. 10.22. Observed dispersion of the Al 10 eV surface loss at large k_x, E_0 = 50 keV /10.51/

A number of papers have appeared which treat this question theoretically /10.58a/. A satisfying solution of the dynamical problem seems not to be achieved. A recent calculation based on a two step model is in agreement with the observation insofar as the disposable parameters (thickness d of the transition film and its electron density $\tilde{n}$) allow one to fit the experiments with reasonable values (d = 4 Å, $\tilde{n}$ = 0.7 n) /10.58/.

Sensitivity against surface changes. The properties of SP's are sensitive to modifications of the surface. Such changes can be 1) a coating of the boundary with a thin film of another dielectric function, 2) a change of the geometry of the boundarylike roughness or grating structure. The geometric changes and their influence on the dispersion relation have been studied with SP's excited by light, see Chap. 10.4. Similar results obtained by ELS are not yet known. In the following the effect of thin coating films on the SP observed with ELS is treated.

Coating films. Coating films on the plasma boundary and their influence on the SP's have been studied extensively, see /10.1c/. From (10.9b) it can be seen that on a plasma surface covered with a thick film ($k_x d \gg 1$) of ε_0 the loss peak appears at $\varepsilon' = -\varepsilon_0$ for large k_x (ε_0 real). This causes a displacement of the peak position to lower energies due to the depolarising effect of the film. A well-known example is the displacement of the surface loss of a clean Al boundary from 10 eV to ∿ 7 eV, if the metal is coated with an oxide film /10.59/. If the coating is not transparent (nonvanishing imaginary part of ε_0) in addition to the displacement of the loss peak an increase of the halfwidth takes place as can be observed on silver films covered with thin carbon films, as can be derived from (10.49). Contrary to the volume loss the surface loss can be nearly suppressed by carbon coating /10.60/.

If the coating consists of another plasma film, e.g., Mg covers an Al film, then in addition to the two SP's $Mg/_{vac}$ and $Al/_{vac}$ an SP at the boundary Mg/Al can be excited. In the case that both metals are free electron gases with $\tau \to \infty$ the equation $\varepsilon' = -\varepsilon_o'$ gives

$$\omega_s^2 = \omega_{s,1}^2 + \omega_{s,o}^2. \qquad (10.57)$$

This has been confirmed in different experiments: In the example Mg/Al we expect with $\hbar\omega_s$ (Mg) = 7 eV and $\hbar\omega_s$ (Al) = 10 eV the SP at 12.9 eV in agreement with the observed value of 12.8 eV /10.50,61/.

The behaviour of two adjacent plasma films, one of them much thinner than the other, is treated in Sect. 10.4.

Thickness dispersion. In considerations above the thickness was assumed to be large ($k_x d \gg 1$). If the condition is not fulfilled as for thin coating films, the frequency of the SP depends strongly on their thickness. This thickness dispersion is displayed in Fig. 10.23: with increasing θ or k_x the SP frequency decreases to the asymptotic value $\omega_p(1+\varepsilon_o)^{-1/2}$, the more rapid with k_x the larger the thickness of the coating. The experiments verified this relation as shown in Fig. 10.24 /10.43/. Here an asymmetric layer system has been investigated: one surface of a clean Al film (surface loss 10 eV) is covered step by step with oxide of different thickness, whereas the other boundary of the Al film is coated with an oxide film of 150 Å thickness (loss value 6.7 eV) serving as substrate for the thin Al

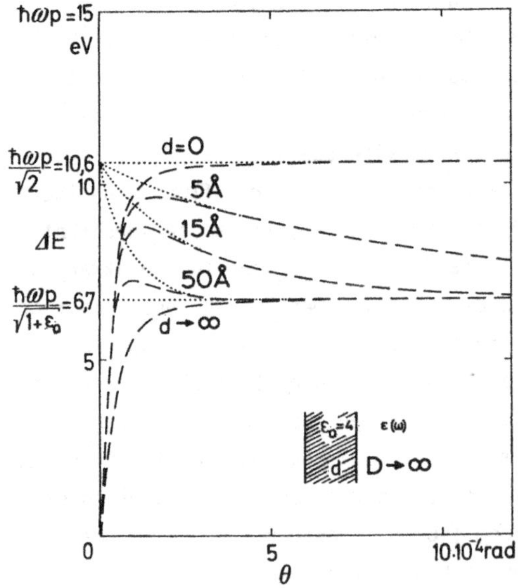

Fig. 10.23. Dispersion curves $\omega(\theta)$ of Al covered with coatings of different thickness $d(\varepsilon_o = 4)$; 50 keV electrons /10.43/. Dotted line: without retardation

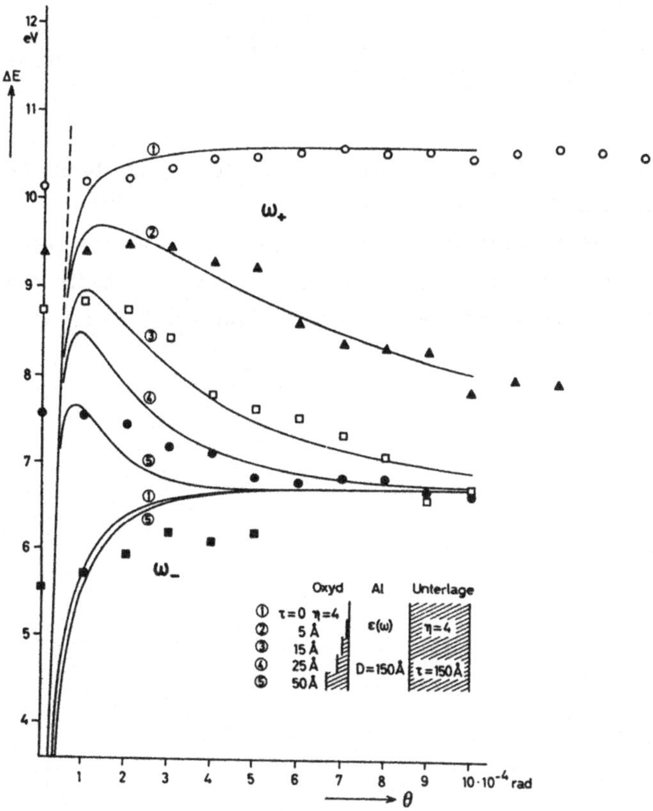

Fig. 10.24. Observed "thickness" dispersion on Al covered with oxide of varying thickness. The upper curves 1-5 indicate the ω^+, the lower curve (filled squares) the ω^- frequency /10.43/

film. Starting the experiment with a clean surface, curve (1), in Fig. 10.24, an asymptotic value of $\hbar\omega^+ = 10$ eV and $\hbar\omega^- = 6.7$ eV, curve (1), is measured. The influence of different thickness is shown by curves (2) to (5). The influence on ω^+ is much stronger than on ω^-. For curve (5) we approach a nearly symmetric system where coupling between the modes at the two surfaces leads to the splitting similar to Fig. 10.21 /10.28/.

If we look at Fig. 10.23 at constant ω the k_x value of the SP displaces with increasing thickness to larger θ values, similar to the curve $m_p = 0$ in Fig. 10.11. This reflects the thickness dispersion.

Nonnormal incidence. The excitation probability of plasmons for nonnormal incidence of the electron beam is given by /10.14,52,53/

$$\frac{\partial^3 W_s}{\partial \Delta E \partial^2 \Omega} = \left(\frac{e}{\hbar \pi v}\right)^2 \cdot k_{el}^2 \cdot \mathrm{Im}\left[\frac{1}{\varepsilon}\frac{d}{\cos\alpha}\frac{1}{q^2} - \frac{2}{\cos\alpha}\frac{(\varepsilon-\varepsilon_0)^2}{\varepsilon\varepsilon_0(\varepsilon+\varepsilon_0)}\frac{|q_s|}{q^4}\right] \tag{10.58}$$

with α the angle between the film normal and the direction of the primary beam and q_s the projection of q onto the boundary, see Fig. 10.25. Comparing (10.58) with (10.44), which gives the excitation probability for normal incidence, we see that in the first term the thickness d is replaced by $d/\cos\alpha$, since the path length for volume excitation increases with α and in the second term the surface term k_x is replaced by q_s.

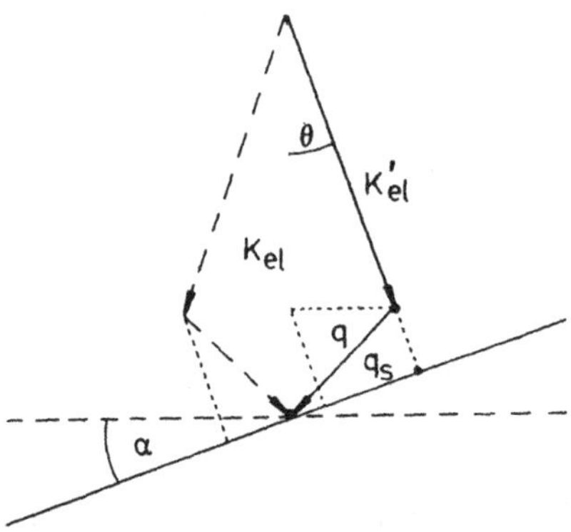

Fig. 10.25. If an electron beam (k_{el}) hits a plasma film at oblique incidence the projection of q onto the film plane leads to different values: Right of the primary beam the projection of $q_\parallel$ and $q_\perp$ are added, left subtracted. $q_\parallel$, $q_\perp$, see Fig. 3.1

As Fig. 10.25 displays we can write for q_s

$$q_s = q_\perp \cos\alpha \pm q_\parallel \sin\alpha = k_{el}\theta \cos\alpha \cdot \left(1 \pm \frac{\theta_{\Delta E}}{\theta} \mathrm{tg}\alpha\right). \tag{10.59}$$

The ± sign is valid whether we have to add (q full line) or to subtract (q dashed line) the projections of $q_\perp$ and $q_\parallel$ onto the surface. Equation (10.58) has two consequences:

a) The loss intensity is different "right" or "left" from the primary beam since q_s does not have the same value. This asymmetry has been observed first by CREUZBURG /10.62/ on Ge and Si surface losses; later on the surface losses of Al /10.63/, Ag /10.64/, and Mg /10.47/ have been studied. This asymmetry effect has been measured with high angular resolution too /10.28/; the results confirmed the calculations rather well.

b) If we perform experiments at grazing incidence of the electron beam, α approaches 90^0, so that in (10.59) the term

$$q_s/\cos\alpha = k_{el}\theta\left(1 \pm \frac{\theta_{\Delta E}}{\theta}\,\mathrm{tg}\alpha\right) \tag{10.60}$$

approaches, /1.1c/, Eq. (10.87)

$$q_s/\cos\alpha = k_{el}\theta_{\Delta E}/\cos\alpha. \tag{10.61}$$

This dependence means that the electrons interact with the plasma boundary a longer time for larger α than for smaller α.

Experiments with thin Ag films transmitted with fast electrons demonstrated that the increase of the intensity of the surface loss (3.6 eV) is well described up to large α ($\sim 80^0$) by $(\cos\alpha)^{-1}$ /10.65/.

Reflection experiments with fast electrons. The considerations deduced for non-normal incidence can explain the results of the reflection experiments especially the dependence on α. Experiments on cleavage surfaces of Si, Ge /10.66/, electrolytically polished Al /10.67/, and liquid metals /10.68/ demonstrated that the loss intensity of SP's compared to that of the volume losses increases rapidly with larger α. Thus transmission at normal incidence and reflection loss spectra differ strongly as has been shown at Si /10.66/. Experiments with 10 keV electrons at grazing incidence up to $\alpha \sim 89^0$ on liquid metals (In, Ga, Al) which have a very smooth surface appropriate for measurements at large α demonstrated the validity of the $(\cos\alpha)^{-1}$ law. Figure 10.26 shows a typical reflection spectrum of liquid In (8.4 eV surface loss) with multiples, but no volume loss.

By changing the angle α the value of W_s, (10.54), can be varied so that the validity of the Poisson distribution of the intensity of the multiple losses has been demonstrated /10.45/.

The $(\cos\alpha)^{-1}$ dependence of the intensity of the surface loss means that the volume loss intensity is nearly suppressed in the reflection spectrum. The latter however can be found back by looking at the angular dependence of the loss: The spectrum of the crystalline Si face at an incidence angle of 85.6^0 of 14 keV electrons in a Bragg spot shows a pronounced surface loss (11 eV) and no volume loss, see Fig. 10.27. This is valid for the center of the diffraction spot, marked with $\theta = 0$. The intensity distribution outside the center for different θ values (in the plane of incidence) however shows that with increasing θ the volume loss intensity (17 eV) comes out of the background and the SP loss intensity is reduced. This behaviour is due to the different dependence on θ (volume intensity θ^{-2}, surface intensity θ^{-3}). The existence of the volume excitation indicates a certain penetration depth of

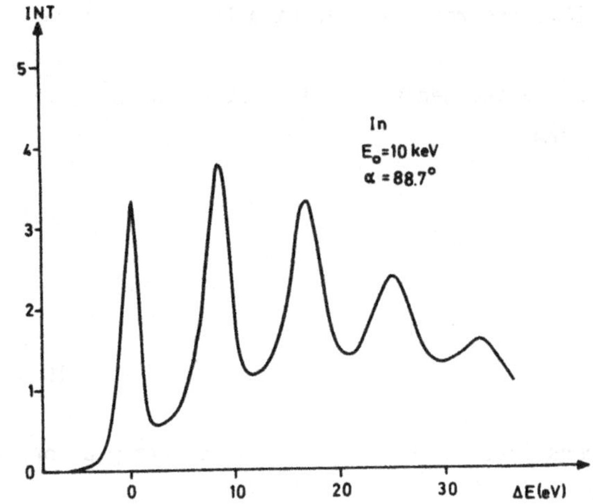

Fig. 10.26. Typical energy loss spectrum of 10 keV electrons reflected at a liquid Indium surface /10.45/

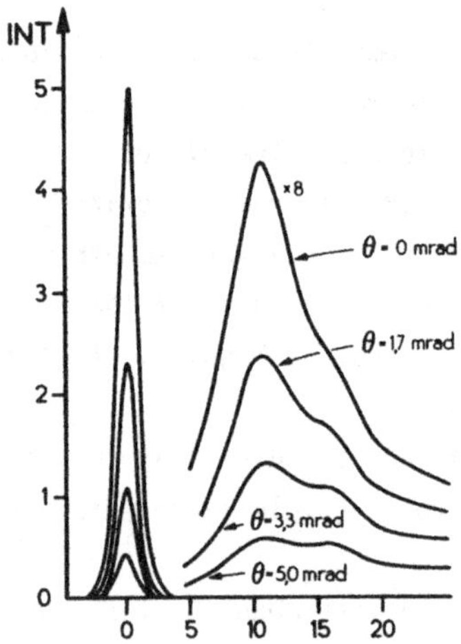

Fig. 10.27. Loss spectrum of a Si monocrystal in reflection at an angle of 85.6° against the normal of the surface. 14.2 keV electrons. The spectrum θ = 0 has been observed in the center of the Bragg spot; the angle θ is measured against the center of the diffraction spot in the plane of incidence /10.45/

the electron into the bulk which has been estimated to approximately 10 Å, which seems reasonable /10.45/.

Experiments have been made with electrons (E_0 = 2 keV) which follow a parabolic trajectory so that they just do not touch the surface /10.69/. This is an interesting approach to study the losses at grazing incidence. The results however are still preliminary.

Reflection experiments with low-energy electrons. The reflection method is a powerful tool when using slow electrons since the electrons are rather sensitive to surface properties due to their strong interaction with the upper layer of the crystal, also at nongrazing incidence. Another advantage of small energy primary electrons (E_0) is the possibility to produce monochromatised electron beams of good intensity. At last rather well-developed experimental equipment is commercially available to start experiments in this field.

This ELS method has found a lot of applications: Surface and volume plasmons together with interband transitions have been observed in agreement with the results of ELS with fast electrons /10.70/. In addition surface states, e.g., on Si (111) crystal faces, have been found in agreement with the results obtained with photoelectrons /10.71/. Of actual importance is the high sensitivity of the ELS to detect small quantities of molecules adsorbed at the surface /10.72,73/ and their vibrations /10.74/.

Losses observed in the specular reflected beam, e.g., from surfaces of Pt (111) and Ni (111) exposed to CO gas, can be attributed to vibrations of the adsorbed molecules of CO against the metal surface /10.74/. The concept is the following: In the electric field between the incident electron and its mirror charge which is perpendicular to the surface and nearly homogeneous at the surface the total dipole moment - the dipole moment of the adsorbed molecule plus the dipole moment of its mirror image - is excited; in a first approximation vibrations perpendicular to the metal surface are thus observed. Pt (111) faces exposed to CO at a temperature of 150 K show peaks at 59 meV (476 cm^{-1}) and at 259 meV (2089 cm^{-1}) (coverage about 0.2 monolayers), Fig. 10.28 /10.74,75/. With increasing coverage further peaks at $\sim$ 47 meV and at 230 meV start to grow, see Fig. 10.28. The smaller energy losses are attributed to the vibration CO-Pt surface, whereas the higher energy losses are explained as being due to the dipole of the stretching mode (C-O vibration) together with its mirror dipole so that the total dipole moment is twice that of the C-O vibration. The CO molecule stands perpendicular with its C atom on the metal surface.

The energy of these modes is dependent on the site at which the C atom touches the metal atoms, here the Pt (111) atoms: adsorption on top sites leads to the losses 59 and 259 meV, whereas the lower energies (47 and 230 meV) are due to the bridge bonded CO molecule. In other experiments the adsorption of atomic hydrogen on W (100) faces has been measured; its surface vibrational modes were detected and its dependence on the reconstruction of the lattice /10.76/. The ELS in combination with the infrared spectroscopy /10.70/ represents an important tool for surface physics.

This technique of slow electrons is also of interest for the physics of plasmons. It allows one to measure the threshold electron energy at which the plasma excita-

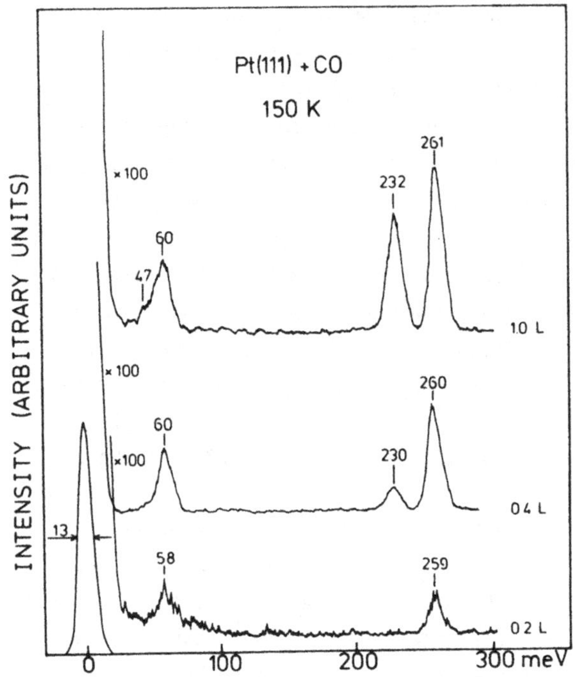

Fig. 10.28. Vibration losses of CO adsorbed on Pt (111). The spectrum CO/Pt changes with increasing CO coverage. The splitting of the modes is due to the adsorption of the CO molecules in different sites of the Pt surface /10.74/

tion starts and the energy dependence of the plasmon loss intensity. First calculations are found in /10.79/ and first experiments are reported /10.80/.

Reflection intensity. In the following some remarks are added concerning the intensity of electrons, inelastically scattered at surfaces, see also /10.46a,78/.

This quantity has been calculated classically as well as with quantum mechanics. In the classical case the electrons are treated as a particle which hits the surface under the angle α measured against the normal. It is now assumed that the excitation probability of surface plasmons for electrons having been reflected at the angle α from the surface is the same as that for electrons having transmitted the surface at the angle α. This probability as function of θ, the angle against the direction of the reflected beam, is given if we observe in the plane of incidence, see (10.58) with $\varepsilon_0 = 1$,

$$\frac{\partial^3 W_s}{\partial \Delta E \partial^2 \Omega} = \left(\frac{e}{\hbar \pi v}\right)^2 k_{el}^2 \frac{8}{\cos\alpha} \frac{q_s}{q^4} \operatorname{Im} \frac{-1}{1+\varepsilon(\omega)} \qquad (10.62)$$

with q_s the projection of q on the surface

$$q_s = \frac{\omega}{v} \sin\alpha + k_{el}\theta \cos\alpha . \tag{10.62a}$$

A similar calculation has been made with quantum mechanics /10.77/. The result for plasmon excitation is the same and has the form

$$\frac{\partial^3 W_s}{\partial \Delta E \partial^2 \Omega} = \frac{2m^2 e^2 v_\perp^2}{\pi} \frac{R^2}{\cos\alpha} \frac{v_\perp^2 q_s}{\left[v_\perp^2 q_s^2 + (\omega - v_\parallel q_s)^2\right]^2} \cdot P(\omega, q_s) \tag{10.62b}$$

$v_\perp$ and $v_\parallel$ are the components of the initial electron velocity perpendicular and parallel to the surface, R is the reflection coefficient. In (10.62b) the reflection coefficient of the process (1) at first reflection and then inelastic process or (2) at first inelastic process and then reflection is set equal. This is identical with the assumption that R is independent on the energy between E_0 and E_0-ΔE. $P(\omega,q_s)$ means a surface analogy of the structure factor $S(\omega,q)$ as discussed in Chap. 3 which comes out as /10.77/

$$P(\omega, q_s) = \frac{2}{\pi} \operatorname{Im} \frac{-1}{1+\varepsilon(\omega, q_s)} . \tag{10.62c}$$

These considerations neglect retardation ($\omega/q_s \ll c$) and spatial dispersion ($q_s \ll 2\pi/a$). Due to (10.62a) the transferred momentum is given by $\hbar q_s$ which is finite for $\theta = 0$, in contrast to (10.44). In order to fulfill $q_s \ll 2\pi/a$, the value of θ has to be small.

Since the denominator of (10.62b) is identical with

$$v_\perp^2 k_{el}^2 (\theta^2 + \theta_{\Delta E}^2) = v_\perp^2 q^2 \tag{10.62d}$$

the quantum mechanical result is identical with (10.62). It shows that the excitation probability of surface plasmons has a maximum of $\theta = 0$. This is equivalent with

$$\omega = q_s v_\parallel \tag{10.62e}$$

as (10.62b) shows. Equation (10.62e) means that the electron velocity along the surface agrees with the phase velocity of the SP, which is a well known description of the resonance phenomenon.

The θ dependence of the excitation probability (10.62) which goes with q_s/q^4 or approximately with θ^{-3} shows that the inelastic intensity is concentrated in the forward direction (the direction of the reflected beam). This explains the rather favourable situation to detect such excitations as just described with ELS in spite of their small probability.

In these considerations the volume effects are neglected; one assumes that the electrons do not penetrate into the bulk, they interact during the specular reflection (or Bragg reflection) with the field of the charge density fluctuations in the crystal and excite the SP. At small $\cos\alpha$ this is a good approximation, but in reality volume effects are observed, see Fig. 10.27. In the classical calculation it has been tried to include such effects, see /10.45/.

The absolute value of the reflectivity R which is dependent on several not well defined surface properties, is unknown. However for comparison with the experimental data the intensity in the inelastic channel with that in the elastic after reflection is divided so that R is not needed.

Measurements with fast electrons ($\sim$ 10 keV) on (liquid) In, Ga, and Al on the surface loss given by $\varepsilon(\omega) = -1$, have demonstrated that the (integrated) probability follows the $\cos^{-1}\alpha$ law and that its multiples obey the Poisson distribution. They show further a dependence on the angle α; the ratio: observed to calculated intensity which has the value 1 at grazing incidence, increases with decreasing angle α to larger values. This can be interpreted as an effect of penetration into the volume /10.45/.

Reflection measurements with slow electrons (E_0 = 50 eV, energy half width 80 meV, $\alpha \sim 80^0$, angular aperture about 1^0) have been reported at cleaved surfaces of GaAs, Ge (111) and Si (111) (2x1) in the energy loss region of some eV /10.163/. This loss structure which stems from the surface function $\mathrm{Im}\{-1/1+\varepsilon(\omega)\}$ - far away from $\varepsilon(\omega) = -1$ - together with the bulk dielectric function $\varepsilon(\omega)$ is roughly in agreement with the calculated one; similar agreement is found in the excitation probability (comparison of the intensity in the inelastic and the elastic channel). It is remarkable that the dielectric theory describes the observations down to such low energies.

In the low energy loss region measured at the Si (111) (2x1) surface a strong excess intensity compared with the calculated one is observed which is ascribed to the excitation of surface states. It is assumed that the transition from the occupied to the empty levels of these subbands due to the dangling bonds produce the energy losses /10.164/. The optical constants of this surface state layer are in approximate agreement with those derived from optical measurements /10.164/.

Spherical and cylindrical surfaces. The plasma boundary can have a spherical or a cylindrical geometry. In these cases the same mathematical procedure as mentioned in Chap. 1 for the plane surface leads to equations which contain implicitly the eigenvalues of the SP.

Spherical geometry. If a sphere of radius r of a material of the dielectric funtion $\varepsilon(\omega)$ is embedded in a dielectric medium of ε_0 we obtain

$$\varepsilon_1(\omega) = -\frac{\ell+1}{\ell}\,\varepsilon_o \tag{10.63}$$

with ℓ integer as condition for the eigenvalue. This is valid for $(\omega_p/c)\cdot(r/\ell) << 1$. Assuming a free electron gas the eigenfrequency comes out as

$$\omega_s = \frac{\omega_p}{(2+1/m)^{1/2}}\,. \tag{10.63a}$$

The lowest and the important mode of this series is the Mie mode

$$\omega_s = \omega_p/\sqrt{3}. \tag{10.64}$$

Its electromagnetic field has a dipole configuration and is thus optically active; it emits and absorbs light. The modes $m > 1$ are radiative too. With $m \to \infty$ ω_s converges to $\omega_p/\sqrt{2}$, the SP of plane surfaces. Similar modes are obtained in spherical cavities or voids interchanging the both dielectric functions

$$\varepsilon_1(\omega) = -\frac{m}{m+1}\,\varepsilon_o. \tag{10.65}$$

Here the interior of the voids inside the metal (ε_1) has the dielectric constant ε_o. In the case that the metal around the sphere can be described by a free electron gas the modes have the eigenfrequencies

$$\omega_s = \frac{\omega_p}{(2+1/m)^{1/2}}\,(1+1/m)^{1/2} \tag{10.66}$$

which lie above the lowest mode $m = \infty$, the plane mode. With $m = 1$ (10.66) gives

$$\omega_s = \frac{p}{\sqrt{3}}\,\sqrt{2} \tag{10.67}$$

which corresponds to the spherical $m = 1$ mode. The $m = 0$ mode has the highest frequency ω_p (monopole breathing mode).

Experiments on spheres of Al /10.142,143/, Ag /10.144/, and Au /10.144/ were successful. Such spheres of a radius of about 100 Å embedded in a thin film of gelatin showed losses which have been identified with the mode $\ell = 1$. For more details and references see /10.1c/.

Small metal particles (spheres) on an insulator-metal layer emit light if a voltage is applied /10.160,161/. For a calculation of a model see /10.162/.

Cylindrical geometry. In contrast to the plane and the spherical geometry the solutions of the cylindrical geometry cannot be separated into p and s modes. The properties of the waves on the cylinder lie somewhat between those of the plane and spherical geometries: The wave propagates along the axis of the cylinder [$\sim \exp i(k_x x - \omega t)$]. k_x is a continuous quantity as it does not exist in the spherical geometry whereas the wave around the cylinder is a standing wave [$\sim \exp(i\, n\theta)$, n integer]. The light line (ll) separates radiative and nonradiative SP. The nonradiative SP on the high frequency side of ll (right of ll) approach the value $\omega_p/\sqrt{2}$ at large k_x whereas near the ll ($k_p r = \omega_p/c \cdot r \sim 1$, 2r diameter of the cylinder) it bends down approaching the ll, not very different from the ω^- mode in Fig. 10.3. For higher n the curves $\omega(k_x)$ approach the line $\omega = \omega_p/\sqrt{2}$ and the ll more and more, for details see /10.145/. The radiative modes on the low frequency side of ll (left of ll) are nearly identical with the line $\omega = \omega_p/\sqrt{2}$ if the parameter $k_p r \ll 1$. With increasing $k_p r$ the mode n = 1 starts at $k_x = 0$ at lower and lower values, whereas the modes with n > 1 approach more and more the line $\omega = \omega_p/\sqrt{2}$. All modes approach the ll asymptotically; for details see /10.146,147/.

There is only an optical observation which contributes results to these calculations. Reflection of light at rather thick ($k_p r \sim 10^3$) cylinders (Al coated quartz fiber of 50 μm diameter) demonstrated minima in the reflected light as function of the angle of incidence. These were explained as resonances of radiative SP's between n = 55 and n = 58 which lie just left of the ll /10.147/.

Detection of nonradiative SP by light. Above we described experiments in which the excitation of SP's is detected by the energy losses of electrons. It is also possible to observe the SP's by the light emitted via coupling with the surface roughness or a grating as explained by Fig. 10.12. This process has been suggested by experiments in which rough silver surfaces /10.81-84/ and Al surfaces /10.85/ irradiated with electrons emitted light of the corresponding SP energy $\hbar\omega_s$ = 3.6 resp. 10 eV. In experiments with Al coated gratings this coupling plasmon-photon is much more transparent as Fig. 10.8 displays: the nonradiative plasmon is coupled via $k_r = 2\pi/a$, a grating constant, with the photon (1 → 2). Looking at the grating irradiated with electrons under different angles of incidence the dispersion relation of SP can be determined along the light line /10.86/. Further experiments on Ag coated gratings studied this coupling process in more detail /10.87/.

It may be noted that in principle light can be emitted from SP's travelling on a smooth surface, if one irradiates one boundary of a thin plasma film, e.g., a silver film, with electrons exciting the SP's whereas the other boundary is covered with a glass prism /10.88/ through which the photons can escape; see prism method, Fig. 10.13. Experiments failed up to now since the electrons passing the silver film produce too much fluorescence light in the glass prism.

Excitation by light. In the following we describe the experimental results of the excitation of nonradiative SP's by using the grating method (a) and the prism method (b).

a) Grating method. A sinusoidally corrugated surface of a plasma produced by depositing a silver film on a photoresist grating is irradiated with p polarised light. Recording the zero order reflectivity as a function of the incidence angle a resonance minimum due to strong absorption of light is observed, if the dispersion relation $\omega(k_x + k_r)$ with $k_r = \pm n(2\pi/a)$ is fulfilled, see Fig. 10.29 /10.89, 90/. Contrary to Fig. 10.29 earlier experiments with commercial gratings had shown

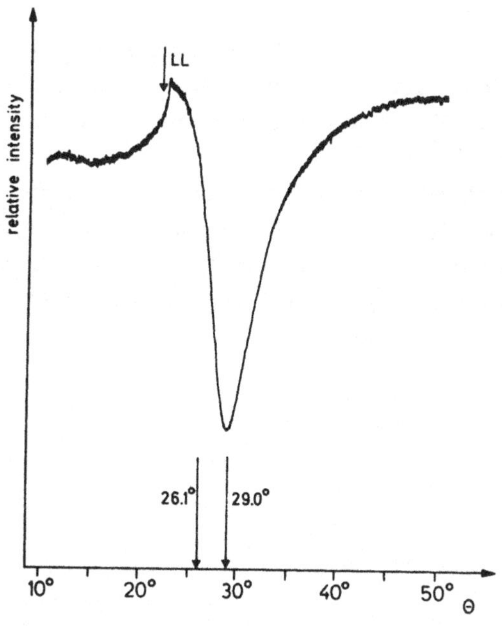

Fig. 10.29. Intensity reflected on a wavy silver film as a function of the angle of incidence θ. The deep minimum indicates the excitation of nonradiative plasmons. Notice the displacement of the measured minimum (29°) with respect to the value calculated for a smooth surface (26.1°) LL means the light line

intensity maxima for the angles (or k_x values) which fulfill the dispersion relation $\omega(k_x+k_r)$ /10.91/. More detailed studies with gratings of nearly sinusoidally corrugated profiles demonstrated a transition of the minima into intensity maxima with increasing amplitudes of the grooves /10.92/. Quantitative evaluation of measurements on nearly sinusoidal gratings indicated that the contribution of higher harmonics to the sinus profile of the grating changes the shape of the reflected intensity rather strongly /10.93/; this explains the observation of intensity peaks instead of minima at gratings of nonsinusoidal profile. The excitation of SP's on gratings and their influence on the diffracted intensity is of more importance for the light spectroscopy which is interested to avoid such variations of the reflected intensity as Fig. 10.29 shows, see /10.94/. As displayed in Fig. 10.28 the measured resonance minimum position differs from the one calculated for smooth surfaces, this point is discussed at the end of this chapter under the topic roughness.

b) Prism method (see Fig. 10.13). As shown above excitation of SP's is possible if the wave vector component $k_x = \frac{\omega}{c}\sqrt{\varepsilon_0}\sin\theta$ of the incident light varies between

$$\frac{\omega}{c} \leq \frac{\omega}{c}\sqrt{\varepsilon_0}\sin\theta \leq \frac{\omega}{c}\sqrt{\varepsilon_0}.$$

In these relatively small limits of k_x near the light line the dispersion relation (10.14) can be studied. The resonance is indicated by a minimum in the totally reflected light, see, e.g., Figs. 10.30 and 10.31, as in the grating method since the SP's are absorbed in the metal rather strongly.

The k_x values of the minima are not identical with the solution of the one boundary system, e.g., Ag/quartz, (10.16,17) and Fig. 10.1, since the plasma film has now a finite thickness d_1, see Fig. 10.13. We have to solve the equation of a two boundary system, e.g., quartz (ε_0)/Ag (ε_1) (thickness d_1)/air (ε_2). Thus we obtain instead of the solution $k_x^{\infty m}$ for $d = \infty$ the value /10.95/

$$k_x^m = k_x^{\infty m} + \Delta k_x^m \ . \tag{10.68}$$

The value of Δk_x^m is given by the relation

$$\Delta k_x^m = \frac{\omega}{c}\,\frac{2}{1+|\varepsilon_1|}\left(\frac{|\varepsilon_1|}{|\varepsilon_1|-1}\right)^{3/2} r_{01}^P(k_x^{\infty m}) \cdot \exp(-2k_x^{\infty m}\, d_1) \tag{10.69}$$

with $r_{01}^P\ (k_x^{\infty m})$ the Fresnel coefficient

$$r_{01}^P\ (k_x^{\infty m}) = \frac{k_{z0}/\varepsilon_0 - k_{z1}/\varepsilon_1}{k_{z0}/\varepsilon_0 + k_{z1}/\varepsilon_1} \ . \tag{10.70}$$

The real part of Δk_x^m produces a displacement of the reflection minimum, the imaginary part of Δk_x^m increases the damping with respect to the solution of (10.16,17) and depends strongly on the thickness d_1. This additional radiation damping is due to the coupling of the SP's with freely propagating electromagnetic waves in ε_0 (e.g., quartz) through the plasma of thickness d_1 which is just the reverse process to the excitation of SP's by light described above.

The reflected intensity $|r^P_{012}|^2$ for the prism arrangement for p polarised light is given by /10.95/

$$R = \frac{|E^P_r|^2}{|E^P_o|^2} = |r^P_{012}|^2 = \frac{r^P_{01} + r^P_{12}\,\exp 2ik_{z1}d_1}{1 + r^P_{02}\cdot r^P_{12}\,\exp 2ik_{z1}d_1}\,, \tag{10.71}$$

see Fig. 10.13. It can be developed around Δk^m_x and for $\exp 2|k_{z1}|d_1 \gg 1$, $|\varepsilon_1'| \gg \varepsilon_1''$ and $|\varepsilon_1'| \gg 1$:

$$R = 1 - \frac{4\gamma_i\gamma_r}{(k_x-k^m_x)^2 + (\gamma_i+\gamma_r)^2} \tag{10.72}$$

$$\gamma_i = \mathrm{Im}(k^{\infty m}_x) \text{ interior damping} \tag{10.73}$$

$$\gamma_r = \mathrm{Im}(\Delta k^m_x) \text{ radiation damping.} \tag{10.74}$$

The depth of the resonance R_{min} is deduced from (10.72)

$$R_{min} = \frac{(\gamma_i-\gamma_r)^2}{(\gamma_i+\gamma_r)^2}\,. \tag{10.75}$$

In our approximation $R_{min} = 0$ is given for

$$\gamma_i = \gamma_r. \tag{10.76}$$

Equation (10.76) which is equivalent to a matching condition known from other optical and electrical problems allows one to derive a thickness d_{opt} /10.95/. This optimum thickness amounts to 500 Å in case of silver and for λ = 500 Å in agreement with observation. Similar behaviour shows the grating coupling: the depth of the reflection minimum in Fig. 10.29 changes with the depth of the grooves h of the grating; it goes through zero at a h_{opt} /10.90/.

This reflection curve (10.71) is very sensitive to changes of the quantities ε_1', ε_1'', and d_1, so that by adapting observed reflection curves to calculated ones the dielectric function together with the thickness of the plasma film can be determined rather exactly /10.95/ and /1.1c/, also under high vacuum conditions /10.100, 101/.

It is interesting that two configurations are possible: 1) The medium ε_1 is a dielectric material (air gap or a solid transparent film) whereas ε_2 is the dielectric function of the plasma, OTTO /10.96/. 2) The medium ε_1 is the plasma and the medium ε_2 air or a dielectric coating, KRETSCHMANN and RAETHER /10.7/. This configuration is not trivial since it has not been recognized at first /10.97/.

Both configurations have their advantages: The first one is suitable for the investigation of crystal surfaces which have to be brought near to the prism base up to about 5000 Å (silver, λ = 5000 Å) which is equal to d_{opt} of this configuration with $\varepsilon_1 = 1$. Thus especially phonon surface polaritons have been studied in detail on different faces of uniaxial and biaxial crystals and their dispersion behaviour /10.98/. The second method is easier to handle and very flexible for other applications /10.1c/. Here a thin plasma film, e.g., of Ag ($\sim$ 500 Å) is vaporised on the base of the prism so that the SP are excited on the exterior side of the film (interface $\varepsilon_1/\varepsilon_0$). The surface properties can be studied as a function of its roughness or of its coating (oxidation or other thin films). In general one does not vaporise the plasma film onto the base of the prism directly, but one uses a thin glass or quartz plate as substrate of the plasma film (ε_1), which can be quickly exchanged.

Both arrangements allow the application of focussed light, so that the reflected light can be observed over a large region in k space /10.99/.

The resonance can also be detected with this configuration by recording the photoelectron yield: Photoelectrons are emitted if the work function is smaller than the energy of the SP. Experiments have been successful on different materials such as Ag /10.100/, Al /10.102/, Li /10.101/, Na /10.101/.

Applications of the prism method to crystal faces are given in Sect. 10.5. Here we describe experiments on the influence of coatings and roughness on the properties of SP's; both fields have been studied in more detail /10.1c/.

Coating film. The influence of a coating film is calculated from the Fresnel coefficient r_{0123} or t_{0123} of a three boundary system, e.g., quartz (ε_0)/metal (ε_1)/coating film (ε_2)/air ($\varepsilon_3 = 1$). In the case of films (ε_2) of small thickness $k_{z2}d_2 \ll 1$ approximations of first /10.95/ and second order /10.103/ have been calculated. In addition to the term Δk_x^m mentioned in (10.68) a second term $\tilde{\Delta k}_x^m$ has to be added (ε_2 complex),

$$\tilde{\Delta k}_x^m = \left(\frac{\omega}{c}\right)^2 d_2 \, \frac{(\varepsilon_2-1)(\varepsilon_2-\varepsilon_1')}{\varepsilon_2} \left(\frac{\varepsilon_1'}{1+\varepsilon_1'}\right)^2 \frac{1}{(1-\varepsilon_1')\sqrt{-\varepsilon_1'}} \tag{10.79}$$

The shift of the resonance given by $\mathrm{Re}(\tilde{\Delta k}_x^m)$ is proportional to the thickness d_2. Its sign for dielectric films ($\varepsilon_2'' \sim 0$): $\varepsilon_2'-1 > 0$ is positive leading to a displacement to greater k_x values. For metal films ($\varepsilon_2' < 0$) the sign depends on whether $\varepsilon_2' < \varepsilon_1'$ or $\varepsilon_2' > \varepsilon_1'$. The imaginary part of $\tilde{\Delta k}_x^m$ describes the broadening of the resonance due to the coating film (ε_2). It is essentially determined by $\mathrm{Re}(\tilde{\Delta k}_x^m)$: if this is > 0 the reflection curve broadens, if < 0 its width is reduced. This is valid also for $\varepsilon_2'' = 0$.

In the following we discuss some experimental results:

a) The plasma film is covered with a thin film without damping ($\varepsilon_2'' = 0$). Coating of silver films with LiF showed a linearly increasing shift of R_{min} with growing d_2 /10.103,104/. Similar effects have been observed on silver and gold films which have been coated with multiples of monolayers of Cd-arachidate each monolayer having a well-defined thickness of 26.8 Å /10.105-107/.

Besides the shift of the resonance an increase of the width of the reflection minimum has been observed in these experiments in spite of $\varepsilon_2'' = 0$. This is due to a change of the spatial electric field distribution with respect to the uncoated plasma film which increases the field in the metal and leads thus to an increased damping /10.103/.

b) If the dielectric films have a finite value of ε_2'' the width of reflection minimum increases rapidly as is observed for carbon coatings on silver /10.103,104/, see Fig. 10.30.

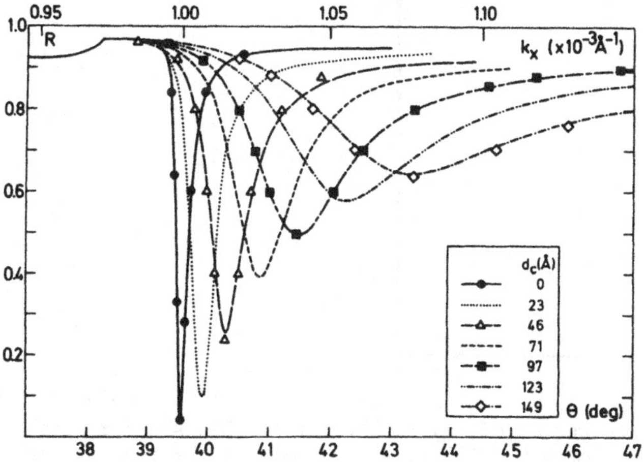

Fig. 10.30. Observed resonance curves of silver films with carbon coatings of different thickness (d_c). Points are the calculated values /10.103/

c) The coating film has a metallic character. Experiments at Ag/Au combinations display the behaviour discussed in connection with (10.79) (Fig. 10.31) /10.108-110/. Whereas gold on a silver substrate leads to a displacement of the reflection minimum to larger θ, see Fig. 10.31, and to stronger damping ($\varepsilon_{Au}' > \varepsilon_{Ag}'$), silver on a gold film has the opposite effect.

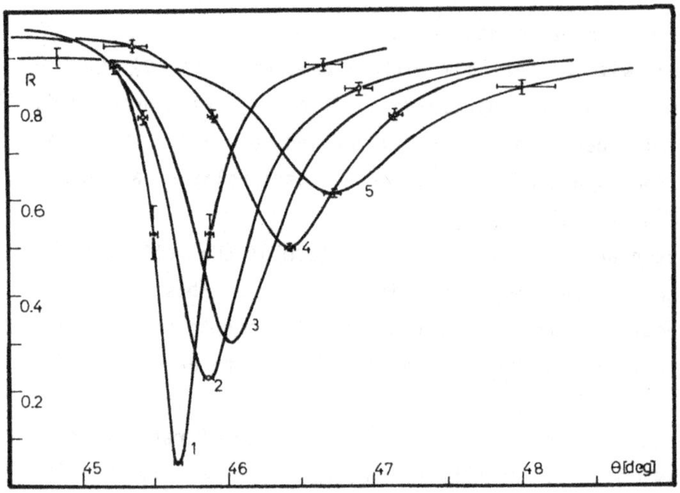

a)

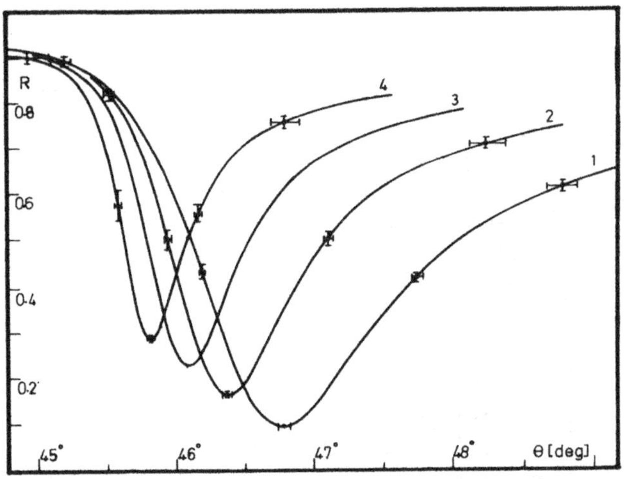

b)

Fig. 10.31a,b. Observed resonance curves of (a) a silver film (570 Å thick) with a thin gold coating; (b) a gold film (570 Å thick) with a thin silver coating. In case (a) the thickness of the gold coating has been (1) 0 Å, (2) 20 Å, (3) 35 Å, (4) 60 Å, (5) 80 Å. Wavelenght $\lambda = 5500$ Å. In case (b) the thickness of the silver coating has been (1) 0 Å, (2) 30 Å, (3) 60 Å, (4) 120 Å /10.109/

d) In general the dielectric function of the coating film has a value $\varepsilon_2(\omega)$ which varies slowly with the frequency. An interesting situation arises if the coating film contains an oscillator with an eigenfrequency ω_0 and a narrow absorption band (small damping). In this case an anomalous behaviour of the shift of the resonance minimum in the frequency region around ω_0 is expected and the width of the reflection minima as well as the depth of the minima shall undergo characteristic, pronounced changes. This has been realized by depositing one monolayer of an organic dye on a thin silver film. The dyes have been either cyanines /10.111,113/ or so-called j aggregates (Scheibe aggregates) which form a rather well-ordered arrangement of dye molecules with a small absorption band /10.112/. These experiments have been made with the dye S 120 (1 octa-decyl-1'-methyl-2.2'cyanine perchlorate); Fig. 10.32, demonstrates the characteristic deformation of the dispersion relation /10.112/.

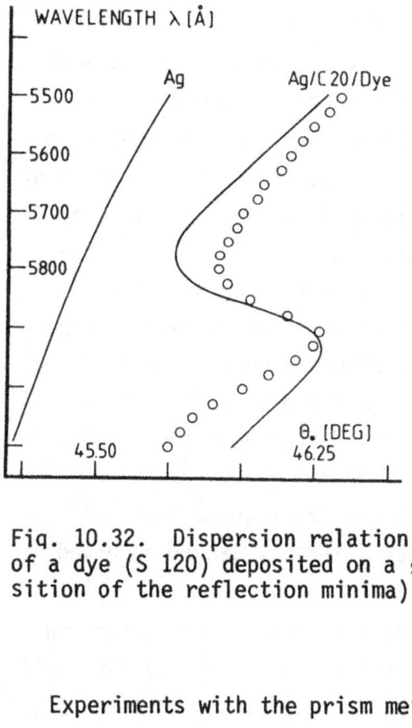

Fig. 10.32. Dispersion relation of a Cd arachidate film covered with a monolayer of a dye (S 120) deposited on a silver substrate (λ_{min} as function of Θ_0, the position of the reflection minima) (o): observed values, full line: calculated /10.112/

Experiments with the prism method have been made on metal (Ag or Au) - electrolyte interfaces, e.g., the presence of an electrolyte 0.5 M $NaClO_4$ on a Ag surface displaces the reflection minimum from 3.85 to 3.03 eV. By applying a (positive) electrode potential the reflection minimum can be displaced to still lower values down to 3.01 eV. This decrease has been interpreted as a change of the free electron concentration in the Ag surface which leads to a decrease of the plasma frequency /10.113a/.

Roughness. An interesting application of the SP's is the study of the roughness, here the geometrical deviation of the profile from a smooth boundary. (The internal roughness is mentioned in Sect. 10.3: "Plasma Resonance Emission"). In the resonance case the strong electromagnetic field in the metal-air boundary produces a high polarisation current in the metal which feeds the peaks of the rough surface acting as antenna, so that they radiate light energy like Hertz dipoles. Its angular distribution is given by the radiation characteristic of the dipole itself (cooperation of the components of the polarisation current perpendicular and parallel to the surface) and further by the spatial correlation of the dipoles which is determined by the structure of the roughness. The theoretical treatment has been given by STERN /10.114/, see also /10.115/. KRÜGER and KRETSCHMANN /10.116,117/ have improved these considerations with the following results: a) The polarisation current producing the scattered light has to be placed into the interior of the plasma $[\varepsilon_1(\omega)]$, if interior roughness (variation of ε_1) is responsible for the scattered light. This means that the field component normal to the surface, E_z, acts with $(1/\varepsilon)\cdot E_z$ in the interior. b) If the surface roughness is responsible for the scattered light, this polarisation current has to be placed into an infinitesimally thin slit at the surface, so that the normal component of the incoming light acts with E_z. This has the consequence that the angular dependence of the scattered light in the two cases is rather different and allows one to discern between interior and surface roughness. Further, in case of interior roughness mainly the light of radiative plasmons ($\varepsilon_1 \cong 0$) will be emitted, whereas surface roughness is characterized by the emission of light from nonradiative plasmons ($|\varepsilon_1| > 1$). These results are important for the evaluation of the experiments with scattered light.

Detailed experiments to determine the roughness of metal surfaces have been performed by using visible light at a silver film, 500 Å thick, vaporised onto a fire polished quartz substrate /10.4,118-122/. The mean roughness height $(\delta^2)^{1/2}$ has been determined to $\sim$ 5 Å a value only valid in the limited region of the roughness spectrum of k_r with $k_r = 2\pi/a$ given by "a" between $\sim$ 3000 and $\sim$ 6000 Å; the wavelength of the scattered light must be comparable with the grating constant a of the Fourier component of the roughness so that diffraction intensity can be observed at $\sin\theta = \lambda/a \leq 1$.

If the roughness of the silver film is increased by underlaying crystalline material on the quartz substrate before depositing the silver the limits of the application of the first-order scattering theory become apparent; this upper limit $(\delta^2)^{1/2}$ is about 20 Å /10.122/.

The geometrical deviation of the plasma surface from a plane one perturbates the SP's so that a change of the dispersion relation is expected. In Fig. 10.29 a displacement of the reflection minimum to larger θ values is seen. Detailed evaluation of the experiments showed that the reflection minimum (or the maximum of the scattered light) is displaced to larger values and its width increases with growing roughness of the surface /10.123-126a/.

To describe the displacement of the reflection minimum to larger θ values multiple scattering (at least double scattering) processes have to be regarded. In a recent paper it has been shown /10.127/ that taking into account multiple scattering the dispersion relation for the plane surface $L = \varepsilon_0 k_{z1} + \varepsilon_1 k_{zo} = 0$ (10.13) has to be replaced by

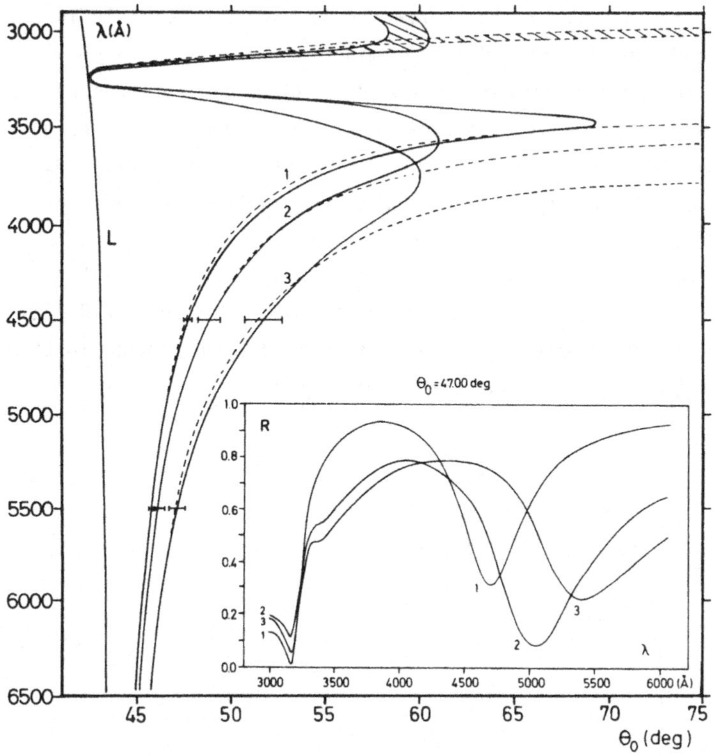

Fig. 10.33. Dispersion relation of silver films of different roughness /10.126a/. λ means the wavelength at which the minimum of reflection R is observed; full line: derived from $R(\theta_0)$ at constant λ. Dashed line: derived from $R(\lambda)$ at constant θ_0. As long as damping is small both curves are identical. Drawing $\lambda(\theta_0)$ instead of $\omega(k_x)$ allows to demonstrate easier the influence of roughness: R_{min} is displaced to larger angles θ_0 with increasing roughness; (1) "smooth" surface: 350 Å Ag vaporised on a fire polished quartz substrate; (2) and (3): underlayer of Ag crystals of different thickness (50 and 100 Å) which increase the roughness of Ag films vaporised on these underlayer. L light line. - The inset shows $R(\lambda)$ at constant θ_0 and the displacement due to increasing roughness and the broadening of the minimum. The backbending in $\lambda(\theta_0)$ and the behaviour of $R(\lambda)$ at shorter wavelengths is discussed in /10.125/

$$L' = L + \text{const.}\ \delta^2 . \tag{10.80}$$

Thus the dispersion relation is changed in such a way that k_x in (10.14) has to be replaced by $k_x + \Delta k_x$ /10.116,117/. Δk_x has a real part which is responsible for the displacement of the reflection minimum to larger θ and an imaginary part which enlarges its width. The term Δk_x is proportional to δ^2.

In principle a comparison of the results of these theoretical considerations with the experimental results /10.123-126/ should be possible but the lack of knowledge of the roughness spectrum at larger k_r values than those mentioned above prevents the evaluation of the formulas. Probably the k_r spectrum in the region of the "a" values comparable with the crystal size at the surface will have a special influence on the displacement and broadening of the reflection minimum.

The multiple scattering theory however can be tested by applying it to the experimental results obtained on sinusoidally corrugated gratings which have a roughness spectrum consisting only of one k_r /10.128-130/. Similarly as on statistically rough surfaces the values of θ_{min} and $\theta_{1/2}$ increase with the amplitudes h of the sinus profile as Fig. 10.34 demonstrates.

The comparison of observed and calculated data shows good agreement /10.131/, see Fig. 10.34. Recent measurements on silver and gold sinusoidal gratings /10.132/ improved the good agreement between experimental and calculated values which in this case have been derived by using the Rayleigh method /10.132/./.

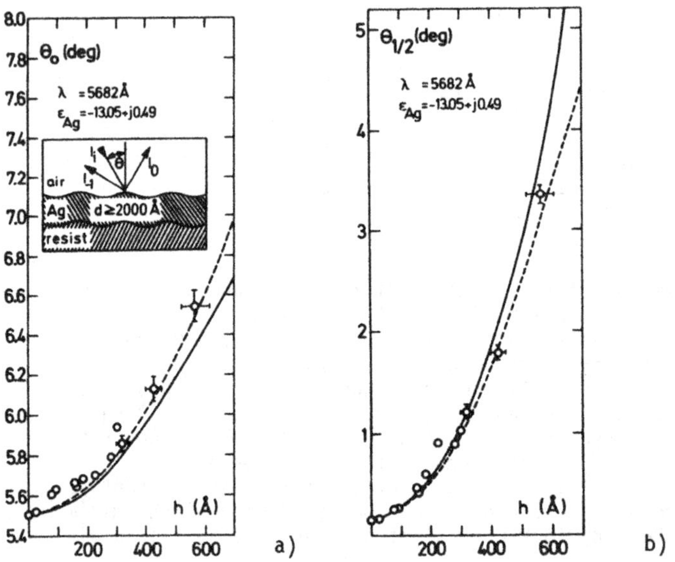

Fig. 10.34a,b. Influence of the groove depth h of a sinusoidal silver grating on the resonance angle θ_o (a) and the halfwidth $\theta_{1/2}$ (b) of the SP resonance. Dots are experimental values; curves are calculated /10.131/

10.5 Surface Waves

Surface plasmons and volume plasmons exist in a free electron gas or in a gas of loosely bound electrons (with a binding energy small with respect to the plasma frequency). Such collective surface phenomena can be found in case of strongly bound electrons too. In this case the behaviour of an electron is classically described by the dielectric function of an oscillator with eigenfrequency ω_0, see Fig. 2.2. If the oscillation is sufficiently undamped the ε' curve passes through negative values with $|\varepsilon'| \geq 1$ (this is not the case in Fig. 2.2). Thus the condition for the existence of surface polarisation waves or polaritons is fulfilled. Negative ε' values prevent light from penetrating the solid and the material shows high, quasimetallic reflection. Surface waves, which exist in this frequency region, may be optically excited by the grating or prism method or by electrons. The oscillator, which is responsible for the negative ε' values, must not be an electronic one, it can also be an ionic oscillator.

There exist several examples of these surface waves:

a) The surface phonons or phonon surface polaritons /10.133/. These modes exist between ω_T and ω_L, the transverse and longitudinal frequency. They have been excited on alkali halide boundaries of LiF, etc., with fast electrons /10.134/ and on ZnO and Si surfaces with slow electrons /10.135/. As their energy lies in the infrared ($\sim$ 50 meV), a good energy resolution was necessary for these experiments. Near the light line (retardation region) the prism method can be applied to detect these phonon surface polaritons optically as experiments on GaP surfaces had demonstrated /10.136,136a/.

Besides these surface phonons bulk phonons are observed with ELS in transmission of thin films of NH_4Cl, NH_4Br, NH_3DBr, ND_4Cl, and ND_4Br /10.137/. The ELS spectra are compared with the optical absorption spectra.

b) The exciton surface polariton. If excitons (electron-hole pairs) exist in the crystal, exciton polarisation waves can propagate in the bulk of the crystal. In case of low damping (ε' values < -1) the condition for the existence of surface modes is fulfilled. These exciton surface polaritons have been observed with the prism method at ZnO surfaces /10.136a/ and on CuBr, CuCl /10.138/. The same phenomenon has been detected on anthracene /10.139/ and on some organic dye crystals /10.140/. For example, measurements on (110) faces of the organic dye CTIP showed even at room temperature a well-pronounced reflection minimum /10.141/.

10.6 2D and 1D Plasmons

The volume plasmons are fluctuations of the density of a three-dimensional (3D) plasma, described by an isotropic dielectric function of a free electron gas. But there exist also two- and one-dimensional plasmas and their corresponding plasmons. Phenomenologically their response is described by strongly anisotropic dielectric functions which have in a certain frequency region negative values in a plane or in one direction, so that quasi 2D and 1D metals result.

Quasi 2D metals. In a 2D electron gas the mobility of the electrons in a (x,y) plane is very high (quasi free electrons) whereas perpendicular to it (z direction) the extension of the gas is comparable with the wavelength of the electrons, so that standing waves or bound states are the consequence. Thus 2D collective motions (longitudinal charge density oscillations of the 2D plasma) can exist in the (x,y) plane which are described as plane waves whereas their field decays exponentially perpendicular to it. The field of the oscillation therefore can be written as

$$\exp i(k_x x + k_y y) - k_z|z| - i\omega t \tag{10.81}$$

with

$$k_z = \sqrt{k_x^2 - \varepsilon_M(\omega/c)^2}, \quad k_y = 0. \tag{10.82}$$

The charges of equal sign lie on rows similarly as for surface plasmons, see Fig. 10.1, and the electric field surrounds these rows.

The dispersion relation of these 2D plasmons can be derived from the dispersion relation of surface plasmons in a thin film of thickness d in the limit $d \to 0$, see (10.9a). The thin film is embedded in a dielectric material with $\varepsilon_M = 1$. Neglecting for the moment retardation one obtains with $k_x d \ll 1$ for the 2D plasma frequency

$$\omega(k_x) = \omega_p/\sqrt{2}\left(1 - e^{-k_x d}\right)^{1/2} = \sqrt{a}\,\sqrt{k_x} \tag{10.83}$$

with

$$a = 1/2\ e^2 n_s/m^* \cdot \tilde{\varepsilon}_0 \tag{10.84}$$

$$\tilde{\varepsilon}_0 = (4\pi \cdot 9 \cdot 10^9)^{-1}\ \text{Coul (Volt m)}^{-1}$$

and

$$n_s = nd \tag{10.85}$$

the electron density per cm^2.

The $k_x^{1/2}$ relation is connected with the fact that the restoring forces determining the eigenfrequency of the oscillations are produced by the electrical fields around the "cylindrical" rows of positive and negative charges. These forces are therefore proportional to $1/\lambda$ (λ wavelength of the 2D plasmons), so that their eigenfrequency becomes dependent on the wave vector, whereas in the 3D case the charges of the volume plasma oscillations are distributed on plane parallel sheets so that the restoring forces (and thus the eigenfrequency) are independent on the distance of the sheets and thus on λ.

The finite velocity of propagation of the electromagnetic field however cannot be neglected at small k_x values. Calculating the response of the 2D electron gas to a longitudinal external electrical field or the 2D Lindhard function, the analog to the 3D dielectric function, see (1.16), the more general dispersion relation is obtained for $\omega/k_x \gg v_{th}$ /10.148/

$$k_x^2 = \varepsilon_M(\omega/c)^2 + (\varepsilon_M\,\omega^2/a)^2 \tag{10.86}$$

if the 2D gas is embedded in a volume of (frequency independent) ε_M; v_{th} means the thermal velocity of the electrons. For small ω and k_x we obtain a linear relation $\omega(k_x)$ approaching the light line $(\omega/c)\sqrt{\varepsilon_M}$.

k_x being larger than $(\omega/c)\sqrt{\varepsilon_M}$ these 2D plasmons are not radiative as the non-radiative surface plasmons in 3D, compare (10.12).

The transition of the linear into the square root region takes place at rather low ω and k_x values, so that the retardation influence disappears rather quickly with increasing k_x.

Experiments with ELS on 2D plasmons have not yet been reported; however other techniques have allowed one to verify these considerations.

1) Electrons on a liquid He surface. Electrons can be deposited on a liquid He surface with a density of 10^6-10^{10} cm^{-2} /10.149/. They are fixed in a potential well of a depth of $\sim$ 1 meV and a spatial extent of about 100 Å to the He/vacuum boundary formed by the image potential [ε(He) = 1.05] and the repulsive barrier of the He surface. The gas is a classical one due to the low density as long as the temperature is higher than about 10^{-3} K. Confining the lateral extension of the electron gas ($\sim$ 1 cm $\cdot$ $\sim$1 cm) standing longitudinal plasma oscillations with small k_x values ($k_x = 2\pi/\lambda \sim 5$ cm^{-1} and more) can be excited by an exterior electric field along the surface. At electron densities of 10^8 cm^{-2} and $\lambda \sim 1$ cm resonance frequencies of about 10^8 Hz have been measured /10.150/. The $\hbar\omega$ of the plasmon is about 10^{-3} meV. If more electrons are spread onto the surface and fixed at it by an additional electric dc field perpendicular to the He surface, the displacement of the resonance frequency with $(n_s)^{1/2}$ to higher values, see (10.83), has been demonstrated /10.150/. Damping of these waves is due to scattering of the electrons at the He vapour (if the temperature of the He is higher than 0.68 K) and to scattering by capillary waves of the surface (ripplons) /10.150/.

As we have just dicussed the 2D electron gas on a liquid He surface behaves as a classsical (nondegenerate) gas, so that its kinetic energy is given by $2 \cdot \frac{1}{2}$ KT. In addition we have to consider the Coulomb interaction $e^2/4\pi\tilde{\varepsilon}_0 r_s$ with $n_s = 1/\pi r_s^2$ which leads to a repulsion of the electrons since screening can be neglected at these low densities. The interesting question arises: can these forces produce a more or less ordered state or an electron crystal ("Wigner lattice") /10.151/, if the kinetic energy is not too high or $E_{pot} > E_{kin}$. Varying

$$\frac{E_{pot}}{E_{kin}} = \frac{e^2\sqrt{\pi}}{4\pi\tilde{\varepsilon}_0 k} \frac{\sqrt{n_s}}{T} \tag{10.87}$$

by changing n_s and the temperature T of the liquid GRIMES and ADAMS /10.152/ succeeded in detecting resonances which are ascribed to the formation of a 2D ordered state of the electrons. For example the transition into the crystal phase at an electron density n_s of $4 \cdot 10^8$ cm^{-2} has been detected by the appearance of resonances if the temperature is reduced below T = 0.4 K; this corresponds to $E_{pot}/E_{kin} = 137$.

2) Electrons in an inversion layer. Another 2D metal can be realized in a Schottky inversion layer on a surface of a semiconductor (nearly free from surface states), e.g., on a Si-SiO_2 metal structure (MOSFET). Values of n_s of 10^{11}-10^{13} cm^{-2} can be produced by an appropriate gate field perpendicular to the surface across the SiO_2 layer. The electrons (or holes) move in a plane in the Si/SiO_2 boundary inside a potential well with high mobility; perpendicular to this plane the carriers are enclosed in a region of $\sim$ 100 Å with the Fermi energy $E_F = \pi h^2 n_s/m^*$ of about 10 meV (at $n_s \sim 10^{12}$ cm^{-2}). Their wavelength amounts to some 10^2 Å.

An excitation of these 2D waves is not possible by light directly, see (10.86), similar to the situation of the nonradiative surface waves. The application of a grating coupler however as described in Sect. 10.4 allows excitation of these waves /10.153/. The inversion layer is irradiated with infrared light through a grating (lattice constant a $\sim$ 5 μ) at normal incidence ($k_x = 0$) of varying frequency, see Fig. 10.8. (In this figure point 2 has to be brought at $k_x = 0$). The radiation has to be polarised perpendicular to the grating. The transmitted light intensity shows peaks of absorption if plasmons are excited, compare with Fig. 10.29. [In the figure 10.29 θ_0 has been varied ($k_r = 2\pi/a$ and λ is constant), whereas here λ resp. the frequency f has been varied ($k_r = 2\pi/a$ and $\theta_0 = 0$ is constant).] The metal electrode at which the gate potential is applied is very thin, so that the system $Si\text{-}SiO_2$ metal is rather transparent.

At values of n_s of some 10^{12} cm^{-2} and wave vector $k_x = k_r = 2\pi/a \sim 10^4$ cm^{-1} the resonance frequency of the incoming ligth lies at $f \sim 10^{12}$ s^{-1} or $\lambda \sim 300$ μ in agreement with the calculated value. The plasmon energy $\hbar\omega$ corresponds to about 1 meV. This relation (10.83) has been controlled by measuring the resonance frequency at different n_s values /10.153/. The dependence of k_x has been demonstrated by using the harmonics nk_r which are present due to the rectangular profile of the grooves of the grating through which the inversion layer is irradiated /10.154/.

The dispersion relation (10.86) has to be corrected due to the presence of the SiO_2 metal layer on the Si surface necessary to apply the gate field. It comes out as /10.155/

$$\omega^2 = ak_x\ (\varepsilon_s + \varepsilon_{ox}\mathrm{ctg\,h}\cdot k_x\cdot D)^{-1} \tag{10.88}$$

with D thickness of the oxide layer and ε_s the dielectric function of the semiconductor. It produces a decrease of the eigenfrequency by the screening of the plasma oscillations by the image charges in the metal electrode. This relation (10.88) is identical with (10.83) if $\varepsilon_{ox} = \varepsilon_M = \varepsilon_{Si} = 1$ and $D \to \infty$.

At higher k_x values a stronger increase of ω^2 than with k_x is expected: a term $\frac{3}{4}\,v_F^2\,k_x^2$ has to be added to $\omega^2 = ak_x$ similarly as the term $\frac{3}{5}\,v_F^2\,q^2$ in the 3D case, see (1.18). The dispersion relation however has not yet been measured at these higher k_x values.

Acoustic 2D plasmons are expected too. Their frequency should decrease to zero with $k_x \to 0$ similar to the acoustic 3D plasmons. In both cases the slope ω/k_x is of the order of the Fermi velocity v_F. For 2D plasmons as explained above the slope is given by the light velocity, see (10.86).

3) Quasi 1D metals. There exist a number of substances which demonstrate along a certain direction a quasi metallic behaviour: the crystal is built up in such a way that its structure can be described as being composed of strands in which the atoms interact strongly, so that electrons are delocalized along these linear rows and behave as quasi free electrons; perpendicular to the strands the interaction is more or less weak. Compounds with such highly anisotropic electrical conductivity are for example KCP, the alkali salt of Pt $(CN)_4$ partially oxidised with Br or Cl, more exactly $K_2[Pt(CN)_4]_2\ Br_{0.3}\cdot 3H_2O$. In this substance exists a special direction in which the Pt atoms have a distance of 2.81 Å, slightly larger than 2.77 Å the distance of the Pt-Pt distance in the metal, so that metallic conductivity in this direction results. Further crystals are TTF-TCNQ, $(SN)_x$ etc., see /10.156/.

The conductivity in the strands of KCP can amount to a few 10^2 $(\Omega\ cm)^{-1}$ at room temperature [Pt metal $1\cdot 10^5$ $(\Omega\ cm)^{-1}$]. perpendicular it is at least 10^3 times lower

/10.156/. Light polarised parallel to the strands shows metallic reflection and a plasma edge is observed in the reflected light at ∿ 2 eV, where $Re(\varepsilon_{\parallel}) = 0$.

ELS experiments have been realized as soon as thin, more or less monocrystalline films of about 1000 Å thickness of these materials could be brought into vacuum without decompositing; KCP for example loses H_2O in vacuum and thus its 1D properties. Successful experiments have been made with TTF, TCNQ /10.157/ and $(SN)_x$ /10.158/ which have been prepared as thin films with 1D direction *in* their surface. Fast electrons which pass such a thin film at normal incidence excite plasmons in the strands. If the energy losses are observed at a finite angle θ or with a finite momentum transfer $\hbar\vec{q}$, $\vec{q}$ is nearly perpendicular to the primary electron beam. Adjusting $\vec{q}$ parallel to the strand direction the loss spectrum is determined by the optical properties of the strand. [With (3.12) one calculates that an angle θ of 0.5 mrad (q ∿ 0.06 $Å^{-1}$) is sufficient to turn q by 88° out of the direction of the primary beam].

The loss spectrum of these substances at room temperature demonstrates a rather strong peak at low energies of about 1-2 eV. In epitaxially grown thin films of $(SN)_x$ this peak lies at 2.48 eV and has a line width of < 0.5 eV. This value corresponds to the energy of the plasma edge measured in the reflected light, so that this excitation can be interpreted as a plasmon peak.

The theoretical interpretation /10.159/ assumes that valence electrons in 1D crystals oscillate in the direction of the strand axis with the plasma frequency $\omega_p = ne^2/m\tilde{\varepsilon}_0$ with n electron density per cm^3 as in 3D. They obey the dispersion relation $\omega(q) = \omega_p + \frac{\hbar}{m}\alpha q^2$ with

$$\alpha = \frac{3}{5}\frac{E_F}{\Delta E} ,$$

see (1.20), valid theoretically at $q \ll q_F$. A further result of the theoretical considerations is the lack of Landau damping since the energy of the collective oscillations always exceeds the energy of single particle excitation in contrast to 3D plasmons, see Fig. 1.2. If the electrons are not free - this is demonstrated by several losses due to interband transitions observed at higher energies of several eV - the dispersion curve mentioned above can change drastically its shape; it can have even a negative slope.

Within this concept the peak of 2.5 eV in $(SN)_x$ corresponds to a free electron plasmon calculated as 6.4 eV; it is observed at 2.5 eV due to the interband transitions at higher energies of 6.6 and 8.7 eV. It behaves thus similar to the loss at 7 eV in graphite (π plasmons). The dispersion of this loss is normal insofar as it has a positive slope like in metals - its energy increases up to 3.4 eV at 0.55 $Å^{-1}$ - and follows the theoretical curve at small q values with a dispersion coefficient α = 0.62. The width of the loss peak increases with q^2 linearly as at 3D volume plasmons, see Chap. 6. This result indicates that in $(SN)_x$ the plasma electrons can be regarded as relatively free. In contrast ELS of TTF - TCNQ had the result that the loss peak which lies for small q at 0.75 eV displaces to 0.55 eV at large wave vectors near the Brillouin zone (~ 1 Å^{-1}), i.e., it has a negative dispersion. The loss is due to a collective excitation, since the real part of the dielectric function $\varepsilon'(\omega)$ goes through zero at this energy and ε'' is small. Apparently the negative dispersion has to be interpreted as due to the influence of the band structure; the latter appears in the loss spectrum as interband transitions at 1.4, 2.1, and 3.5 eV which demonstrate no significant anisotropy.

The strong anisotropy of the quasi free electron properties can be observed by turning the film around the electron beam so that $\vec{q}$ makes an angle $\bar{\theta}$ with the strand axis *in* the film surface. It is found experimentally that the loss energy decreases; theoretically it should follow a relation $\omega(\bar{\theta}) = \omega_{p\_} \cos\bar{\theta}$ /10.159/. The observation verifies the decrease: in $(SN)_x$ it is less than $\cos\bar{\theta}$, in TCNQ and KCP (optical experiments) it follows the $\cos\bar{\theta}$ relation /10.158/. Investigating this $\bar{\theta}$ dependence up to 90^0 the dielectric behaviour perpendicular to the strands ($\varepsilon_{\perp}$) is recognized: In $(SN)_x$ the 2.5 eV loss (at $\bar{\theta} = 0$) is displaced to lower values and a small dispersion persists. This is interpreted as: $(SN)_x$ resembles more a very anisotropic 3D metal than a 1D metal; in contrast TTF - TCNQ has a nearly frequency independent, positive dielectric constant $\varepsilon_{\perp}$ indicating an insulator, so the model of a 1D metal is a better description for this material.

11. Different Electron Energy Loss Spectrometers

In the following a short description of electron loss analysers for different electron energies shall be given together with some characteristic data.

1) Description of an electron loss spectrometer for medium energies (50–100 keV), see Fig. 11.1. An electron beam leaving a tungsten cathode, focussed by a Wehnelt cylinder and passing the anode, hits a thin foil (object, see Fig. 11.1). Here the electrons can be scattered inelastically by an angle θ. By an electrostatic beam deflector (D) they are bent back in the direction of the primary beam and enter the loss analyser which spreads out the energy spectrum.

In most cases the cathode is a Steigerwald type which yields under normal heating conditions an electron beam of about 0.5 eV energy width. Installation of a mono-

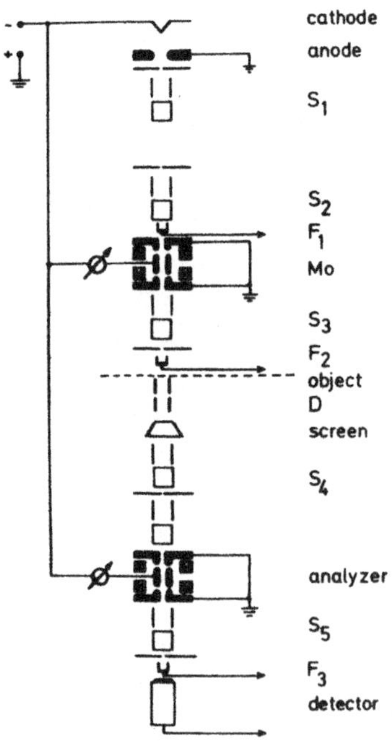

Fig. 11.1. Scheme of an electron loss spectrometer, described in the text /11.7,8/

chromator (Mo) reduces the energy halfwidth, see below. The vacuum is supposed to be around or better than 10^{-6} Torr. In the case that higher vacua are available, better than 10^{-10} Torr, a field emission cathode can be used which produces beams of higher brilliance /11.1,2/ and in addition the energy width of the emitted electrons is smaller; values of 2-300 meV at a current of 0.3 μA have been measured /11.3/. The problem of the actual technical development is to have stable beams and long living cathodes, which seems to be solved to some extent.

The electrons are adjusted to maximum intensity through the diaphragms (30-150 μm) by electrostatic deflection systems (S_1-S_5). The usual angular aperture of the beam on the specimen amounts to about $0.2 \cdot 10^{-3}$ rad (60 keV electrons) or 0.02 $Å^{-1}$. A Faraday cage (F2) allows to measure the current just before the specimen (∿ 0.1 μA).

Special electron optics allows to reduce the angular aperture to values of $8 \cdot 10^{-6}$ rad (75 keV electrons) or $q = 1.2 \cdot 10^{-3}$ $Å^{-1}$ and an energy resolution of some 100 meV. This is of interest, e.g., for the dispersion at very low q values /11.4/.

The specimen is enclosed in a cooled (with liquid nitrogen) metal box to reduce the contamination of the specimen under the influence of the electron beam. The films can also be cooled down to He temperature to solidify, e.g., noble gases /11.5/. Gases can be studied by installing a cell which is filled with vapours or gases from outside.

A fluorescence screen allows one to control the beam during the adjustment to maximum intensity before starting with the registration of energy losses. The electrons scattered by an angle θ are brought back into the axis of the spectrometer by a double plane parallel plate system (D) just behind the specimen /11.6/. Figure 11.2 is a schematic drawing. Its angular calibration is performed by the elec-

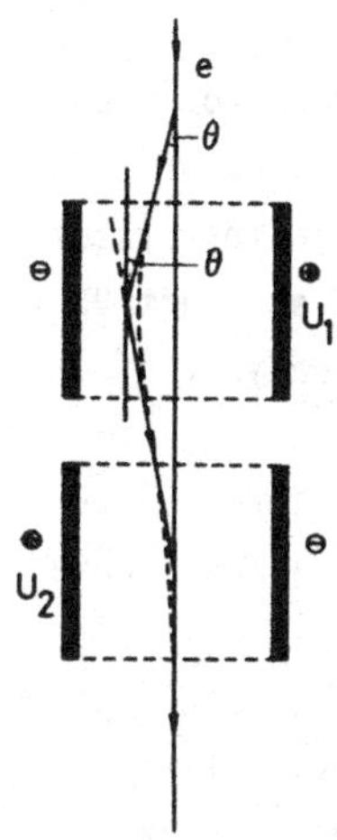

Fig. 11.2. Double plate deflector D in Fig. 11.1. The electrons scattered in a collision process by the angle θ are bent back into the primary direction by the two condensors with opposite field directions /11.6/

tron diffraction maxima of different substances. The electron diffraction pattern can be observed on the screen; this possibility is of special value for the study of monocrystalline films to adjust the right orientation of the wave vector $\vec{q}$ in the crystal.

The electrons after having passed the object enter the slit of the loss analyser. The most common type of analyser is that in which an electrostatic field deflects the electrons according to their energies. We find the parallel plate, the cylindrical, and the spherical analysers; in the latter a stigmatic image of the source is produced. The cylindrical mirror analyser is often applied in low-energy electron studies and is commercially available.

It is characteristic of these analysers that the electrons have to be retarded from some 50 keV to less than 100 eV before entering the analysing field, if a good energy resolution has to be achieved. For example, spherical analyser with an entrance diaphragm of 10^{-2} cm, a radius of 10 cm, and electron velocity in the condensor of 20 eV has an energy resolution of about 10 meV. Figure 11.3 shows an analyser with the retarding electric field; the electrons are detected by an open multiplier and registered with a multichannel analyser or more perfect data processing.

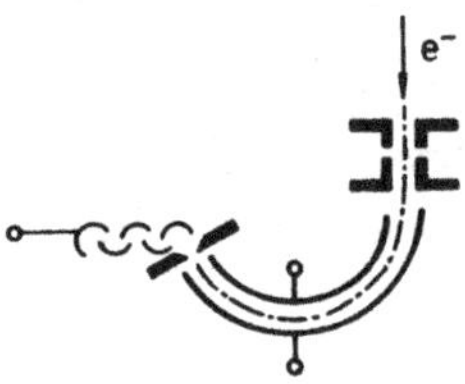

Fig. 11.3. Scheme of an electrostatic deflection analyser. The electrons (e^-) are retarded before they enter the deflection condensor. The exit slit determines the energy width measured with the multiplier

In contrast to these analysers the HARTL lens /11.7/ (see below) does not need a special exterior decelerating field: The electrons leave the filter lens of this system with almost their initial energy of 50 keV and get into the detector system (plastic scintillator and photomultiplier) which has an efficiency of at least 95 %. This monochromatising system can be used as a very stable operating analyser of high resolution too /11.8/.

We find a similar situation in the Möllenstedt analyser which is a three-electrode system of high chromatic aberration. This analyser is often used; for a more detailed description see /11.9/.

2) Energy analyser with high resolution. The reviewed results demonstrate that high-energy resolution is needed for a number of problems as the determination of natural linewidth, excitonic phenomena, the fine structure of the spectrum, and before all phonon effects which have an excitation energy about 10^3 times smaller than plasmons. Several proposals have been made to cut out of the energy distribution of the primary electrons a small region without losing too much intensity. Here in general a compromise between intensity and energy resolution is necessary. It is not yet possible to concentrate the whole number of electrons leaving the cathode into a small energy range.

a) Hartl filter. A rather simple way of building a monochromatising system which is easy to construct and yields rather good intensity is the filter lens (three--electrode system /11.10/), see Fig. 11.4. The outer electrodes have the potential of the anode, the middle electrode lies on the filter voltage U_{FL}. If this is equal to the potential of the cathode U_0, the electrons cannot pass the lens. The focus length of the lens is a strongly oscillating function for small values of U_0-U_F, as Fig. 11.5 demonstrates. Here 1/f is plotted schematically against U_0-U_f. The

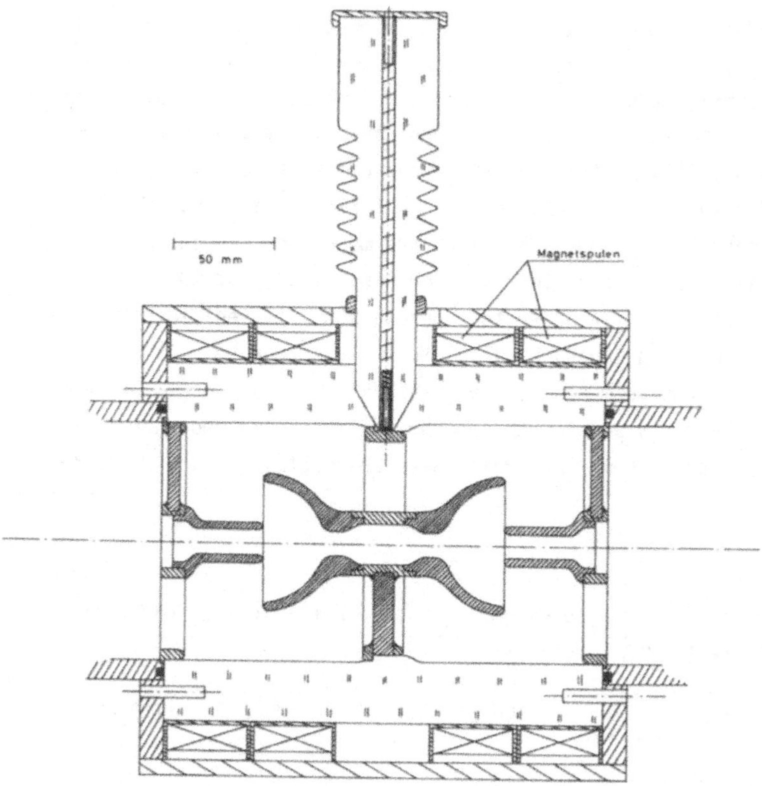

Fig. 11.4. Schematic drawing of the HARTL filter lens /11.7/

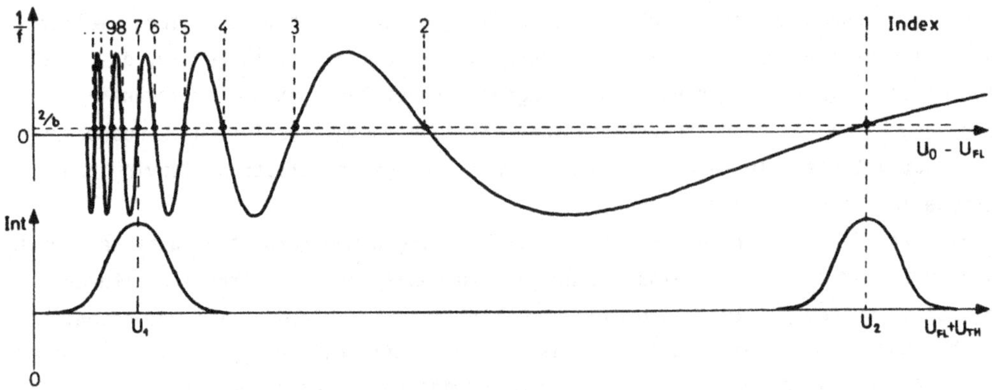

Fig. 11.5. The reciprocal value of the focus length as function of U_0-U_{FL} (U_0 voltage of the outer electrodes, U_{FL} voltage of the inner electrode) /11.11/

line 2/b indicates the distance of the diaphragm S_3 from the lens. If the oscillating function 1/f crosses the line 2/b, the electrons are focussed near the diaphragm and pass the diaphragm (30-100 μm). The energy of the electrons coming from the cathode and entering the lens extends in general over about 0.5 eV (see the peaks in Fig. 11.5 below left), so that several focus points superpose; each of them has an energy width which is proportional to $[d(1/f)dE]^{-1}$ and thus small near the zeros of 1/f. The more U_0-U_{FL} approaches zero the more foci lie inside the energy width of the primary electrons (left side of Fig. 11.5), and the smaller is the energy width in one focus. In order to separate the different focus points a magnetic field (about some ten Gauss) parallel to the electron beam is applied. By a slight inclination of the beam against the lens axis the foci could be separated. By changing U_{FL} the focus point with appropriate intensity and energy width can be chosen. Table 11.1 shows some data which have been obtained by a HARTL filter as analyser /11.7/.

Table 11.1. Energy resolution of a spectrometer with a Hartl lens at different currents. Angular aperture 0.02 $Å^{-1}$, 50 keV

Current at the	object	400 pA	200 pA	110 pA
	multiplier	46 pA	33 pA	17 pA
Energy resolution		150 meV	110 meV	70 meV

b) Wien filter. In the Wien filter crossed electric (E) and magnetic (H) fields are applied /11.6,12,13/. The fast electrons coming from the cathode are decelerated in the first lens in Fig. 11.6 and enter the filter. Here they move perpendicular to the fields E and B ($E \perp B$), so that only electrons with the velocity v = E/B can pass the filter. Having left the condensor the electrons are accelerated again in the lens behind the filter and can be observed. Figure 11.6 shows schematically the

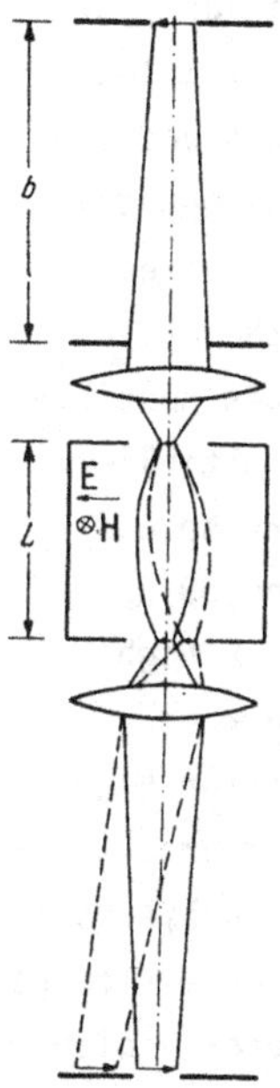

Fig. 11.6. Path of the electrons in a Wien filter /11.12/. The electric field lies in the plane of drawing, the magnetic field lies perpendicular to it

path of the rays. The plates of the deflecting system are perpendicular to the plane of drawing. The energy resolution of the electron beam reaches about 10 meV at a current of about 10^{-12} A at the detector.

Combining two Wien filters, one as monochromator, the other one as analyser, one has a high-energy resolution spectrometer. Thus the losses produced by the excitation of surface phonons have been photographed /11.12/.

This monochromator can also be installed in an electron microscope /11.14,15/ since the electrons remain in the axis of the instrument. Energy resolution of the primary beam of ∿ 40 meV (25 keV electrons) has been reported /11.15/.

3) Loss spectrometer with electrostatic deflection analyser. An energy loss spectrometer operating at 300 keV with an electrostatic deflector (cylindrical condensor) as monochromator and as analysator has been described in /11.16/. A rough scheme of the apparatus is given in Fig. 11.7; its length is about 5 m. The electrons leaving

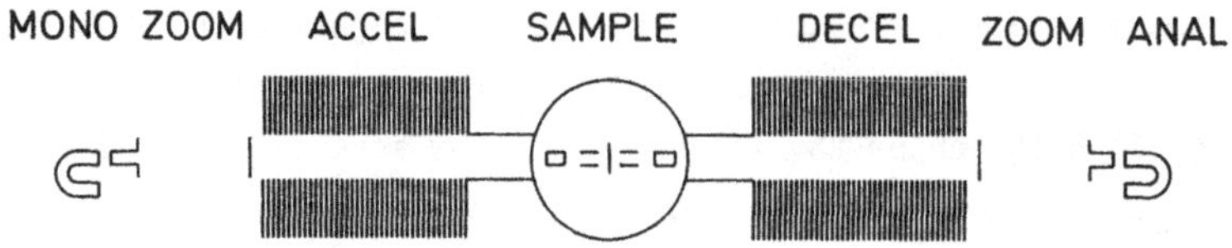

Fig. 11.7. Scheme of an electron loss spectrometer for 300 keV electrons

the cathode which is on ground potential like the detection system - similar potential distribution has been applied in a 50 keV spectrometer successfully /11.16a/ - are monochromatised (Mono), then accelerated to 300 keV in the zoom lens and the accelerating tube and focussed on the sample. These high-energy electrons penetrate thicker films than 50 keV electrons; the films are therefore easier to prepare. Having passed the sample the electrons are decelerated and analysed in the cylindrical deflection system, constructed similar to the monochromator. It was possible to get an energy resolution of 80-120 meV at a current in the electron multiplier of $2\text{-}10 \cdot 10^{-9}$ A. The angular resolution amounts to about 0.1 Å which corresponds to 0.3 mrad (300 keV electrons). This value is equivalent to 0.8 mrad for 50 keV electrons; it is a relatively large aperture which explains the relatively high intensity at the detector. (The irradiated area of the object amounts to 2 mm^2.)

4) Loss spectrometer for high-energy electrons. The loss spectroscopy for high energies, 1 MeV and more, has been developed rapidly recently. This is important for the study of thick specimens of several 10^3 Å and more. The high-energy analysers mostly use magnetic fields /11.17/. Instead of an electrostatic deflection system as above a magnetic $\pi/2$ prism is used in a 1 MeV apparatus to produce the energy dispersion /11.18/. As Fig. 4.3 shows an energy resolution of approximately 2-3 eV has been reached. This limit is due to the ripple of the high-voltage source which is about $\pm$ 1.5 eV. More recently an Ω energy analyser has been studied which deflects the electron beam back into the direction of the primary beam and can be used as energy filter in a microscope too /11.19/, see Figs. 11.8 and 11.9 /11.20/. It is built up from three magnetic systems which disperse the inelastic electrons; a slit C' selects an energy range of electrons chosen for the detection of the loss or for the image formation of interest in the electron microscopic study. The energy resolution is the same as that mentioned above.

An important progress of this Ω analyser is the possibility to study energy filtered images in the region of high electron energy.

5) Loss spectroscopy with low-energy electrons. As already mentioned, Chap. 3, the loss spectroscopy with low-energy electrons is performed with the reflection method. The scheme of a spectrometer is shown in Fig. 11.10. The electrons are monochromatised by a 127° cylindrical electrostatic deflector which has a good

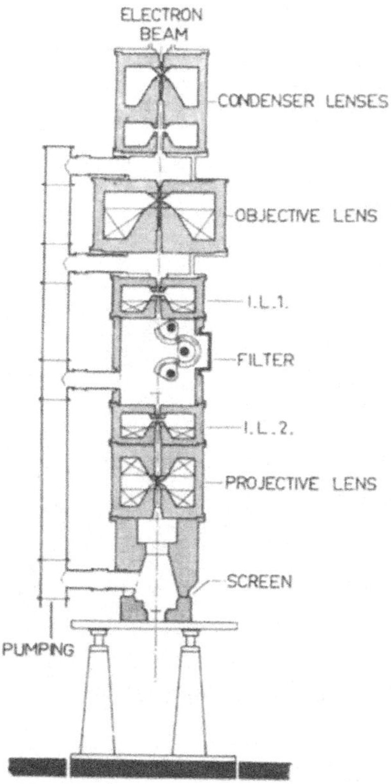

Fig. 11.8. Schematic drawing of a high-voltage microscope /11.19/

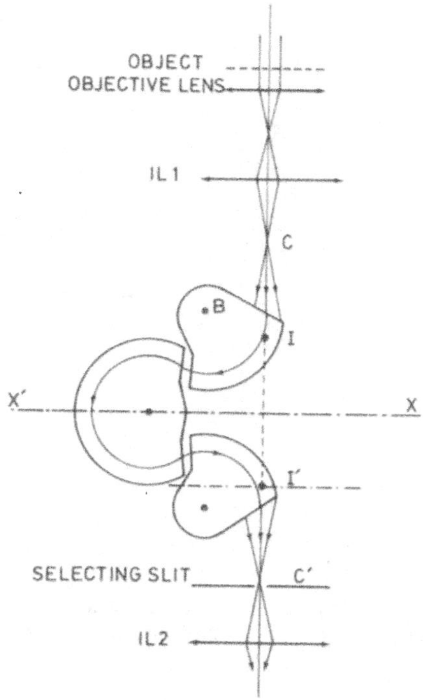

Fig. 11.9. Ω filter built of three magnetic prisms. The electrons enter the first (upper) magnetic prism and are deviated in the clockwise direction. The next prism bends them into the third prism (below) which brings the electrons back into the primary direction /11.20/

energy resolution, so that losses of about 10 meV and less are separated. The angle of incidence of the electrons on the crystal can in general be strongly varied. The scattered electrons are detected in an energy analyser of the same construction as the monochromatiser. The current which is measured at the crystal with a Faraday cage has a value of about 10^{-9} A at an energy halfwidth of the electron beam of about 10 meV; the current in the detector amounts to some 10^{-11} A (E_0=40 eV) /11.21/.

The beam width after the reflection of the slow electrons using the specularly reflected beam is limited by the quality of the reflecting surface and depends on the angle of incidence. The latter dependence is also observed with fast electrons /11.22/. A value of the beam aperture of about 1.5° is often used. This is equivalent to $\sim$ 0.6 mrad at 50 keV electrons or 0.08 $Å^{-1}$.

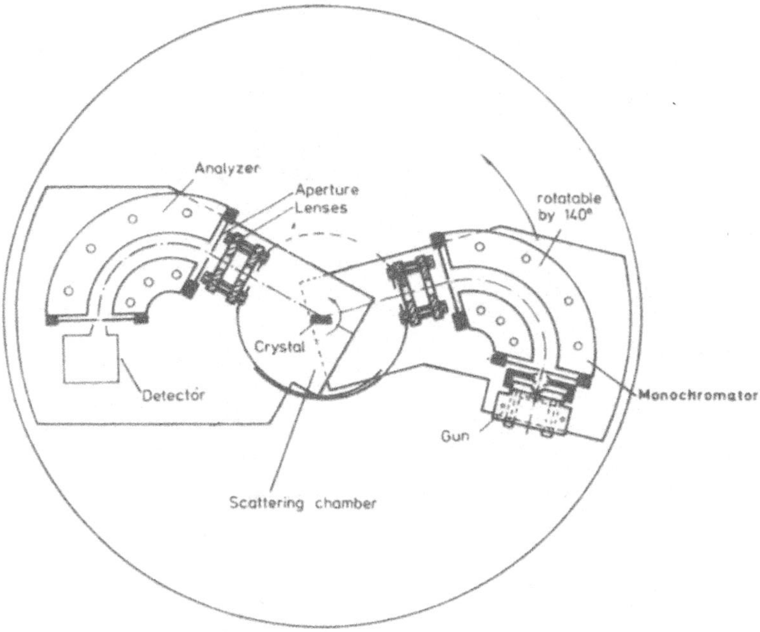

Fig. 11.10. Schematic drawing of a high-resolution spectrometer for ELS of low-energy electrons in reflection /11.21/

Notes Added in Proof

1. Chap. 3: *Plasmon energy gain in the Auger spectrum.* The Auger emission spectrum of a metal, e.g., of Al and Mg excited by X-rays shows besides volume and surface energy losses gains of $n\hbar\omega_p$, where n are whole numbers /M. Suleman, E.B. Pattinson: J. Phys. F: Metal Phys. 1, L21 (1971); L.H. Jenkins, M.F. Chung: Surf. Sci. 26, 151 (1971)/. An interpretation of this observation /see A.D. Matthew, C.M.K. Watts: Phys. Lett. A 37, 239 (1971)/ is given as follows: If the X-ray photon ejects the electron from the core level, the remaining in the very first moment unscreened hole interacts with the free electron plasma. The system hole-plasma can be left in an excited state with an energy of the level plus quanta $n\hbar\omega_p$. This energy can be transferred to an Auger electron so that one observes on the high energy side of the Auger line additional peaks belonging to $n\hbar\omega_p$. This interpretation has been confirmed experimentally /J.G. Fuggle, L.M. Watson, D.J. Fabian, S. Afrossman: J. Phys. F: Metal Phys. 5, 375 (1975); Th.M. van Attekum, J.M. Trooster: J. Phys. F: Metal Phys. 8 L169 (1978); S.A. Abo-Namous, P.T. Andrews, C.E. Johnson: J. Phys. F: Metal Phys. 9, 61 (1979); J.G. Fuggle, R. Lässer: J. Vac. Sci. Tech. (in press) and theoretically I.W. Gadzuck: J. Elect. Spectr. 11, 355 (1977)/. Satellites in the L Auger spectra found in early work are certainly due to double ionisation satellites /H. Loefgren, L. Walldén: Sol. State Comm. 12, 19 (1973); I.G. Fuggle: J. Phys. F: 7, L 81 (1977)/.

2. Sect. 7.2: *Dispersion at large q values.* The measurements of the dispersion of the inelastic peak in the single particle continuum ($q > q_c$) showed the result that the values of $\Delta E(q)$ lie essentially below the calculated (RPA) curve. Recently a similar result has been obtained in measurements at monocrystalline Si /unpublished data of C.H. Chen, A.E. Meixner, B. Kincaid (Bell Telephone). I have to thank the authors for the data/.

3. Chap. 8: *Structure at the absorption edge.* See /S.E.G. Slusky, S.E. Schnatterly, P.C. Gibbons: Phys. Rev. B 20, 379 (1979)/.

4. Sect. 10.4: *Reflection experiments.* The intensity of ineleastically scattered electrons for volume and surface plasmons in transmission as well as in reflection is concentrated in the forward direction around the primary beam (in reflection:

around the specularly reflected beam). This corresponds to the long-range plasmon field interacting with the primary electron. If the q width of the beam amounts to about 2 $k_{el} \cdot (\Delta E/2E_0) = 2\pi/A$ or about some 10^{-2} Å^{-1} (with the usual data of E_0 and ΔE) A corresponds to some 100 Å spatial extension. The experiments with slow electrons have shown a similar distribution behaviour of the energy losses due to the vibrations of the adsorbates as, e.g., CO on Pt or Ni. This indicates that these inelastic processes are produced by the long range field of the fluctuating dipole of the vibrating adsorbate, perpendicular to the surface. Recent ELS experiments on W(100) surfaces and Pt(111) covered with atomic hydrogen showed, in addition to excitations of the type just described, losses with much broader angular intensity distributions. The intensity of these losses which are ascribed to vibrations parallel to the surface could be measured outside the specularly reflected beam up to angles of 30° (angular width of the no loss beam ~ 1°). It is nearly constant or has a weak maximum outside the reflected beam. Further it is a strong function of the energy of the incident electrons and represents a resonant behaviour. These properties indicate a short range field configuration ("impact scattering") as known from scattering by gas molecules, however, with characteristic differences due to the neighbourhood of the surface. /W. Ho, R.F. Willis, E.W. Plummer: Phys. Rev. Lett. 40, 1463 (1978) [W(100) substrate] and A.M. Baro, H. Ibach, H.D. Bruchmann: Surf. Sci. (in press) [Pt(111) substrate]; I have to thank Prof. H. Ibach for the preprint of this paper/.

5. Sect. 10.4: *Surface plasmons, roughness, light emission from inelastic electron tunneling.* Light emission can be observed visually from a metal-insulator-metal junction, e.g., Al-Al_2O_3 (~ 30 Å thick) - metal M (Ag, Au or In), if a voltage of a few eV is applied. Electrons tunneling through the junction excite nonradiative surface plasmons at the metal-insulator interface. Roughening of the electrodes by etching or by CaF_2 underlayer couple the plasmons with photons so that the junction area emits light /J. Lambe, S.L. McCarthy: Phys. Rev. Lett. 37, 923 (1976); S.L. McCarthy, J. Lambe: Appl. Phys. Lett. 30, 427 (1977)/. If the metal film M of the three layer system described above is made of small metal particles (100-200 Å diameter) - which is a special form of roughness - light is emitted too as described in /10.160,161/. This system has been considered theoretically /10.162/.

6. Sect. 10.6: *2D plasmons.* Recent literature on 2D plasmons see Yamada Conference (II) on "Electronic Properties of Two-Dimensional Systems", Lake Yamanaka, Japan (1979).

7. Sect. 10.6: *1D plasmons.* Linear metallic conductivity has been obtained too in doped polyacetylene $(CH)_x$ /C.K. Chiang et al.: Phys. Rev. Lett. 39, 1098 (1977)/.

8. Chap. 11: *Electron Energy Loss Spectrometer.* A similar spectrometer as described in Sect. 11.3 has been built with greater emphasis on beam intensity, using a low work function oxide cathode with an energy resolution of 300-500 meV without monochromator and with poorer angular resolution (~ 0.2 $Å^{-1}$). Currents at the sample of 10^{-6} A (beam diameter ~ 0.2 mm) and of 10 nA at the detector entrance (multiplier) have been obtained. These properties are appropriate for measurements at high momentum transfer and of deep core excitations. I have to thank G.S. Brown and A.E. Meixner (Bell Telephone) for providing me with these data of their spectrometer.

References

1.1 Review articles on this subject:
Experimental work:
a) L. Marton, L.B. Leder, H. Mendlowitz: "Characteristic Energy Losses of Electrons in Solids", in *Advances in Electronics and Electron Physics VII* (New York 1955) where earlier literature is collected
b) J. Geiger: *Elektronen und Festkörperphysik* (Vieweg 1968)
c) H. Raether: "Solid State Excitations by Electrons", in *Springer Tracts in Modern Physics*, Vol. 38 (Springer, Berlin, Heidelberg, New York 1965) pp. 85-155
d) J. Daniels, C. von Festenberg, H. Raether, K. Zeppenfeld: "Optical Constants of Solids by Electron Spectroscopy", in *Springer Tracts in Modern Physics*, Vol. 54 (Springer, Berlin, Heidelberg, New York 1970) pp. 77-135
e) K.D. Sevier: *Low Energy Spectrometry* (Wiley Interscience, New York 1972). (This book contains a chapter on high energy electron loss spectroscopy, pp. 278-322)
f) Theoretical work:
D. Pines, Ph. Nozières: *The Theory of Quantum Liquids* (Benjamin 1966)
g) P.M. Platzman, P.A. Wolff: "Waves and Interactions in Solid State Plasmas", in *Solid State Physics*, Suppl. 13 (Academic Press 1973)
h) R. Becker, F. Sauter: *Theorie der Elektrizität*, Vol. 3, p. 393 (Teubner 1969)

1.2 H. Raether: "Surface Plasma Oscillations and their Applications, in *Physics of Thin Films*, Vol. 9 (Academic Press 1977) pp. 145-261

1.3 P. Nozières, D. Pines: Phys. Rev. 111, 442 (1958)

1.4 D. Mermin: Phys. Rev. B 1, 2362 (1970)

2.1 J. Hubbard: Proc. Roy. Soc. A 243, 336 (1957)

2.1a H.G. Wessjohann, Z. Phys. 269 269 (1974)

2.2 F. Bassani, G. Pastori Paravicini: *Electronic States and Optical Transitions in Solids* (Pergamon Press 1975)

2.3 D. Brust, I.C. Phillips, F. Bassani: Phys. Rev. Lett. 9, 94 (1962); D. Brust: Phys. Rev. A 134, 1337 (1964)

2.4 H. Ehrenreich, H.R. Philipp, B. Segall: Phys. Rev. 132, 1918 (1963)

2.5 F. Bassani, G. Pastori Paravicini: Nuovo Cimento B 50, 95 (1967)

2.6 H. Ehrenreich, M.H. Cohen: Phys. Rev. 115, 786 (1959)

2.7 S.L. Adler: Phys. Rev. 126, 413 (1962)

2.8 N. Wiser: Phys. Rev. 129, 62 (1963)

2.9 W.M. Saslow, G.F. Reiter: Phys. Rev. B 7, 2995 (1973)

2.10 K.C. Pandey, P.M. Platzman, P. Eisenberger: Phys. Rev. B 9, 5046 (1974)

2.11 R. Girlanda, M. Parrinello, E. Tosatti: Phys. Rev. Lett. 36, 1386 (1976)

3.1 G. Ruthemann: Naturwissenschaften 30, 145 (1942); Ann. Phys. 2, 133 (1948)

3.2 P. Nozières, D. Pines: Phys. Rev. 113, 1254 (1959); D. Pines, P. Nozières: *The Theory of Quantum Liquids* (Benjamin 1966)

3.3 P. Schmüser: Z. Phys. 180, 105 (1964)

3.4 P. Zacharias: J. Phys. C 7, L26 (1974)

3.5 R.A. Ferrell: Phys. Rev. 107, 450 (1957)

3.6 J. Geiger: Z. Naturforsch. 17a, 696 (1962)

3.7 H. Venghaus: Z. Phys. 239, 289 (1970)

3.7a H.G. Wessjohann: Thesis, Hamburg (1976); Z. Phys. 269, 269 (1974); Phys. Status Solidi (b) 77, 535 (1976)

3.8 J.C. Ashley, R.H. Ritchie: Phys. Status Solidi 38, 425 (1970)
3.9 D.L. Misell, A.J. Atkins: J. Phys. C 4, L81 (1971)
3.10 J.C.H. Spence, A.E.C. Spargo: Phys. Rev. Lett. 26, 895 (1971)
3.10a J. Ruwalds, A.K. Rajapogal, J. Carballo, G.S. Grest: Phys. Rev. Lett. 36, 274 (1976)
3.11 E. Rudberg: Phys. Rev. 50, 138 (1936)
3.12 R. Feder, N. Müller, D. Wolf: Z. Phys. B 28, 265 (1977)
3.12a Z. Vager, D.S. Gemmell: Phys. Rev. Lett. 37, 1352 (1976);
3.12b V.N. Neelavathi, R.H. Ritchie, W. Brandt: Phys. Rev. Lett. 33, 302 (1974)
3.13 G. Priftis, A. Theodossiu, K. Alexopoulos: Phys. Rev. Lett. A 27, 577 (1968)
3.14 A. Tanokura, N. Hirota, T. Suzuki J. Phys. Soc. Jpn. 27, 515 (1969)
3.15 P. Eisenberger, P.M. Platzman, K.C. Panday: Phys. Rev. Lett. 31, 311 (1973)
3.16 D.M. Miliotis: Phys. Rev. B 3, 701 (1971)
3.16a F. Sauter: Z. Phys. 203, 488 (1967);
F. Forstmann: Z. Phys. 203, 495 (1967)
3.17 A.R. Melnyk, M.J. Harrison: Phys. Rev. Lett. 21, 85 (1968); Phys. Rev. B 2, 835 (1970)
3.18 K. Kliewer, R. Fuchs: Phys. Rev. B 185, 905 (1969)
3.19 I. Lindau, P.O. Nilsson: Phys. Lett. A 31, 352 (1970)
3.20 M. Anderegg, B. Feuerbacher, B. Fitton: Phys. Rev. Lett. 27, 1565 (1971)
4.1 J.N. Hodgson: *Optical Absorption and Dispersion in Solids* (London, 1970)
4.1a H.J. Hinz: Unpublished work (1978)
4.2 M. Creuzburg: Z. Phys. 196, 433 (1966)
4.3 H.J. Hinz, H. Raether: Thin Solid Films 58, 281 (1979)
4.4 M. Creuzburg, H. Raether: Solid State Commun. 2, 345 (1964)
4.5 C. Wehenkel: J. Phys. (Paris) 36, 199 (1975);
C. Wehenkel, B. Gauthé: Phys. Status Solidi (b) 64, 515 (1974)
4.6 D. Misell, A.F. Jones: J. Phys. A 2, 540 (1969)
4.7 J. Daniels: J. Phys. 227, 234 (1969)
4.8 T. Grove: Ultramicroscopy 1, 15 (1975)
4.9 D. Pines: Phys. Rev. 92, 626 (1953)
4.10 J.C. Ashley, R.H. Ritchie: Phys. Status Solidi 40, 623 (1970)
4.11 P. Wienhold: Z. Phys. 208, 313 (1968)
4.12 N. Swanson, C.I. Powell: Phys. Rev. 145, 195 (1966)
4.13 A.W. Blackstock, R.H. Ritchie, R. Birkhoff: Phys. Rev. 100, 1078 (1955)
4.14 M. Horstmann, G. Meyer: Z. Phys. 164, 21 (1961)
4.15 J.Ph. Perez, J. Sevely, B. Jouffrey: Phys. Rev. A 16, 1061 (1977)
4.16 W. Brünger, W. Menz: Z. Phys. 184, 271 (1965)
4.17 L. Marton, J.A. Simpson, H.A. Fowler, N. Swanson: Phys. Rev. 126, 182 (1962)
4.18 B. Jouffrey, I. Sevely: Rev. Phys. Appl. 11, 101 '1976)
4.19 I.O.H. Spence, A.E.C. Spargo: Phys. Rev. Lett. 26, 895 (1971)
4.20 J. Schilling: Z. Phys. B 25, 61 (1976)
4.21 M. Creuzburg, H. Dimigen: Z. Phys. 174, 24 (1963)
4.22 R. Burge, D. Misell: Philos. Mag. 18, 251 (1968)
4.23 H. Kanter: Phys. Rev. B 1, 2357 (1970)
4.24 M. Isaacson: J. Chem. Phys. 56, 1803 (1972)
4.25 J.J. Ritsko, L.J. Brillson, R.W. Bigelow, T.J. Fabish: J. Chem. Phys. 69, 3931 (1978)
5.1 P. Keil: Z. Phys. 214, 251 (1968)
5.2 D.M. Roessler: Private communication
5.3 J. Daniels, P. Krüger: Phys. Status Solidi (b) 43, 659 (1971)
5.4 S. Zimmermann: J. Phys. C: Solid State Phys. 9, 2643 (1976)
5.5 H. Finkenrath, H. Köhler: Z. Angew. Phys. 19, 404 (1965)
5.6 C. Wehenkel: J. Phys. (Paris) 36, 199 (1975)
5.7 J. Daniels: Z. Phys. 227, 234 (1969)
5.8 L.R. Canfield, G. Hass, W.R. Hunter: J. Phys. (Paris) 25, 124 (1964)
5.9 D. Beaglehole: Proc. Phys. Soc. 85, 1007 (1965)
5.10 J. Daniels, C. von Festenberg, H. Raether, K. Zeppenfeld: "Optical Constants of Solids by Electron Spectroscopy", in *Springer Tracts in Modern Physics*, Vol. 54 (Springer, Berlin, Heidelberg, New York 1970) p. 101
5.11 K. Zeppenfeld, Thesis, Hamburg (1969)

5.12 H. Venghaus: Phys. Status Solidi (b) 71, 609 (1975)
5.13 E.A. Taft, H.R. Philipp: Phys. Rev. A 138, 197 (1965)
5.14 H.J. Hinz, H. Venghaus: Proc. IVth Intern. Conf. Vac. UV Rad., Hamburg 1974 (Pergamon-Vieweg 1974)
5.15 H. Venghaus: Z. Phys. 239, 289 (1970)
5.16 G. Meyer: Z. Phys. 148, 61 (1957)
5.17 A. Imbusch, H. Niedrig: Phys. Lett. A 32, 375 (1970)
5.18 J. Schilling: Z. Phys. B 25, 61 (1976)
5.19 C.J. Powell: Phys. Rev. 175, 972 (1968)
5.20 K. Zeppenfeld: Z. Phys. 223, 32 (1969)
5.21 M. Cardona, W. Gudat, E.E. Koch, M. Skibowski, B. Sonntag, P.Y. Yu: Phys. Rev. Lett. 25, 659 (1970)
5.22 H. Philipp, H. Ehrenreich: Phys. Rev. 131, 2016 (1963)
5.23 M. Creuzburg: Z. Phys. 196, 433 (1966)
5.24 M. Piacentini, C.G. Olson, D.W. Lynch: Phys. Rev. Lett. 35, 1658 (1975)
5.25 I.R. Fields, P.C. Gibbons, S.E. Schnatterly: Phys. Rev. Lett. 38, 430 (1977)
5.25a J.J. Ritsko, M.J. Rice: Phys. Rev. Lett. 42, 666 (1979)
5.26 M. Creuzburg: Z. Phys. 194, 211 (1966)
5.27 F. Pradal, C. Gout, B. Lahaye: Solid State Commun. 4, 119 (1966)
5.28 U. Killat: J. Phys. C: Solid State 7, 2396 (1974)
5.29 J. Hainfeld, M. Isaacson: Ultramicroscopy 3, 87 (1978)
5.30 P. Zacharias, K.L. Kliever: Solid State Commun. 18, 23 (1976)
6.1 G. Meyer:Z. Phys. 148, 61 (1957)
6.2 C. Kunz: Z. Phys. 167, 53 (1962)
6.3 T. Kloos: Z. Phys. 265, 225 (1973)
6.4 K.J. Krane: J. Phys. F: Metal Phys. 8, 2133 (1978)
6.5 T. Aiyama, K. Yada: J. Phys. Soc. Jpn. 38, 1357 (1975)
6.6 K. Zeppenfeld, Z. Phys. 223, 32 (1969)
6.7 D. Miliotis: Phys. Rev. B 3, 701 (1971)
6.8 P. Eisenberger, P.M. Platzman, K.C. Pandy: Phys. Rev. Lett. 30, 311 (1973)
6.9 C.H. Chen: J. Phys. C: Solid State Phys. 9, L321 (1976)
6.10 K. Zeppenfeld: Z. Phys. 243, 229 (1971)
6.11 C. von Festenberg: Z. Phys. 207, 47 (1967)
6.12 C. Kunz: Z. Phys. 167, 53 (1962)
6.13 T. Aiyama, K. Yada: J. Phys. Soc. Jpn. 36, 1554 (1974)
6.14 P. Zacharias: Thesis, Hamburg (1974)
6.15 V. Krishan, R.H. Ritchie: Phys. Rev. Lett. 24, 1117 (1970)
6.16 P. Nozières, D. Pines: Phys. Rev. 113, 1254 (1959)
6.17 St.G. Louie, J.R. Chelikowsky, M.L. Cohen: Phys. Rev. Lett. 34, 155 (1975)
6.18 H.J. Hinz, H. Raether: Thin Solid Films 58, 281 (1979)
6.19 H.R. Philipp, H. Ehrenreich: Phys. Rev. 131, 2016 (1963)
6.20 G. Paasch: Phys. Status Solidi 38, K123 (1970)
6.21 M.J. Hasegawa: Phys. Soc. Jpn. 31, 649 (1971)
6.22 T. Kloos: Thesis, Hamburg (1973)
6.23 C. Kunz: Z. Phys. 196, 311 (1966)
6.24 K. Sturm: Z. Phys. B 25, 247 (1976)
6.25 P.C. Gibbons, S.E. Schnatterly, I.I. Ritsko, I.R. Fields: Phys. Rev. B 13, 2451 (1976)
6.26 J. Bösenberg: Z. Phys. 218, 282 (1969)
6.27 K. Sturm: Phys. Rev. Lett. 40, 1599 (1978)
6.28 C.H. Chen: J. Phys. C: Sol. State Phys. 9, L321 (1976)
6.29 H. Bross: J. Phys. F: Metal Phys. 8, 2631 (1978)
6.30 P. Zacharias: Z. Phys. 256, 92 (1972)
6.31 K. Sturm: Z. Phys. B 28, 1 (1977)
6.32 P. Zacharias: J. Phys. F: Metal Phys. 5, 645 (1975)
6.33 D.F. Dubois: Ann. Phys. (NY) 8, 24 (1959)
6.34 B.W. Ninham, C.I. Powell, N. Swanson: Phys. Rev. 145, 209 (1966)
6.35 D.F. Dubois, M.G. Kivelson: Phys. Rev. 186, 409 (1969)
6.36 M.J. Hasegawa, W. Watabe: J. Phys. Soc. Jpn. 27, 1393 (1969)
7.1 H. Watanabe: J. Phys. Soc. Jpn 11, 112 (1956)
7.2 G. Meyer: Z. Phys. 148, 61 (1957)

7.3 T. Kloos: Z. Phys. 265, 225 (1973)
7.4 G. Priftis: Phys. Rev. B 2, 54 (1970)
7.5 A. Tanokura, N. Hirota, T. Suzuki: J. Phys. Soc. Jpn. 27, 515 (1969); 28, 1382 (1970)
7.6 N.G. Alexandropoulos: J. Phys. Soc. Jpn. 31, 1790 (1971)
7.7 P.C. Gibbons, S.E. Schnatterly, J.J. Ritsko, I.R. Fields: Phys. Rev. B 13, 2415 (1976)
7.8 H.I. Höhberger, A. Otto, E. Petri: Solid State Commun. 16, 175 (1975)
7.9 K.J. Krane: J. Phys. F: Metal Phys. 8, 2133 (1978)
7.10 M. Anderegg, B. Feuerbacher, B. Fitton: Phys. Rev. Lett. 27, 1565 (1971)
7.11 K. Zeppenfeld: Z. Phys. 211, 391 (1968)
7.12 K. Zeppenfeld: Opt. Commun. 1, 119 (1969); Z. Phys. 243, 229 (1971)
7.13 H. Venghaus: Phys. Status Solidi (b) 66, 145 (1974)
7.14 P. Eisenberger, P.M. Platzman, K.C. Pandy: Phys. Rev. Lett. 30, 311 (1973)
7.15 R. Manzke, J. Phys. C: Solid State Phys. 11, L349 (1978)
7.16 I. Stiebling: Unpublished work (1978)
7.17 H.J. Hinz: Thesis, Hamburg (1978)
7.18 J. Pyrlik: Unpublished work (1978)
7.19 P. Zacharias, K.L. Kliewer: Solid State Commun. 18, 23 (1976)
7.20 A. Otto, E. Petri: Solid State Commun. 20, 823 (1976)
7.21 I. Lindau, P.O. Nilsson: Phys. Scr. 3, 87 (1971)
7.21a M. Schlüter: Z. Phys. 250, 87 (1972)
7.21b B. Bernert,P. Zacharias: Z. Phys. 241, 205 (1972)
7.22 M. Creuzburg: J. Phys. Soc. Jpn. 20, 1745 (1965)
7.23 M. Urner-Wille, H. Raether: Phys. Lett. A 58, 265 (1976)
7.24a U. Büchner: Phys. Stat. Solidi (b) 81, 227 (1977)
7.24b U. Büchner: Phys. Status Solidi (b) 83, 493 (1977)
7.25 J.P. Walther, M.L. Cohen: Phys. Rev. B 5, 3101 (1972)
7.25a M.S. Hague, K.L. Kliewer: Phys. Rev. B 7, 2416 (1973)
7.26 H. Bross: Phys. Lett. A 64, 418 (1978); J. Phys. F: Metal Phys. 8, 2631 (1978)
7.27 J. Hubbard: Proc. R. Soc. (London) A 243, 336 (1957)
7.28 P. Nozières, D. Pines: Phys. Rev. 111, 442 (1958)
7.29 V.P. Silin:: JETP Soviet. Phys. [Engl. Transl.] 7, 538 (1958)
7.29a L. Hedin, S. Lundquist: Solid State Phys. 23, 172 (1969)
7.30 K.S. Singwi, M.P. Tosi, R.H. Land, A. Sjölander: Phys. Rev. 176, 589 (1968)
7.31 G. Priftis, A. Theodossiou, K. Alexopoulos: Phys. Lett. A 27, 577 (1968)
7.32 D.M. Miliotis: Phys. Rev. B 3, 701 (1971)
7.33 P. Eisenberger, P.M. Platzman, K.C. Pandy: Phys. Rev. Lett. 31, 311 (1973)
7.34 N.G. Alexandropoulos: J. Phys. Soc. Jpn. 31, 1790 (1971)
7.35 D. Marinos, D. Miliotis: Phys. Status Solidi (b) 68, K133 (1975)
7.36 P. Zacharias: Plenarvortrag 37. Physikertagung Hamburg 1973, pp. 405-410; J. Phys. C: Solid State 7, L26 (1974); J. Phys. F 5, 645 (1975)
7.37 H.J. Höhberger, A. Otto, E. Petri: Solid State Commun. 16, 175 (1975)
7.38 P.E. Batson, C.H. Chen, J. Silcox: Phys. Rev. Lett. 37, 937 (1976)
7.39 J. Stiebling, H. Raether: Phys. Rev. Lett. 40, 1293 (1978)
7.40 P.C. Gibbons, S.E. Schnatterly, J.J. Ritsko, R. Fields: Phys. Rev. B 13, 2451 (1976)
7.41 K.L. Kliewer, H. Raether: Phys. Rev. Lett. 30, 971 (1973); J. Phys. C: Solid State 7, 689 (1974)
7.42 P.M. Platzmann, P. Eisenberger: Phys. Rev. Lett. 33, 152 (1974); Solid State Commun. 14, 1 (1974)
7.43 W. Schülke, W. Lautner: Phys. Status Solidi (b) 66, 211 (1974)
7.44 G. Mukhopadhyay, R.K. Kalia, K.S. Singwi: Phys. Rev. Lett. 34, 950 (1975)
7.45 P. Vashista, K.S. Singwi: Phys. Rev. B 6, 875 (1972)
7.46 A.K. Gupta, P.K. Aravind, K.S. Singwi: Solid State Commun. 26, 49 (1978)
7.46a G.D. Priftis, J. Boviatsis, A. Bradis: In press
7.47 M. Urner-Wille: J. Phys. D: Appl. Phys. 40, 49 (1977)
7.48 E. Petri, A. Otto: Phys. Rev. Lett. 34, 1283 (1975)
7.49 F. Bassani, G.P. Paravicini: Nuovo Cimento B 50, 95 (1967)
7.50 A. Otto, E. Petri: Solid State Commun. 20, 823 (1976)

7.51 S.P. Singhal: Phys. Rev. B 12, 564 (1975); Phys. Rev. B 14, 2352 (1976)
7.51a C.H. Chen: Phys. Status Solidi (b) 83, 347 (1977)
7.52 P. Keil: Z. Phys. 214, 251 (1968)
7.53 J. Daniels, P. Krüger: Phys. Status Solidi (b) 43, 659 (1971)
7.54 M. Creuzburg: Z. Phys. 196, 433 (1966); M. Creuzburg, H. Raether: Solid State Commun. 2, 175 (1964)
7.55 J. Hinz: Thesis, Hamburg (1978)
7.56 J. Pyrlik, H. Raether: Phys. Lett. A 63, 355 (1977)
7.57 J. Pyrlik: Solid State Commun. 26, 717 (1978); Z. Phys. B 31, 369 (1978)
8.1 G. Ruthemann: Naturwissenschaften 29, 648 (1941); 30, 145 (1942); Ann. Phys. 2, 135 (1948)
8.2 S. Nomura, H. Todokoro, T. Komoda: J. Electron. Microsc. 26, 277 (1977)
8.3 J. Hillier, R.F. Baker: J. Appl. Phys. 15, 663 (1944); J. Hillier: Phys. Rev. 64, 318 (1943)
8.4 H. Watanabe: J. Appl. Phys. Jpn. 3, 804 (1964)
8.5 N. Swanson, C.J. Powell: Phys. Rev. 167, 592 (1968)
8.6 D.B. Wittry, R.F. Ferrier, V.E. Cosslett: Brit. J. Appl. Phys. 2, 1867 (1969)
8.7 C. Colliex, B. Jouffrey: C. R. 720, 673 (1970); Philos. Mag. 25, 491 (1972)
8.8 A.V. Crewe: Science 154, 729 (1966); 168, 3937 (1970)
8.9 A.V. Crewe, M. Isaacson, D. Johnson: Rev. Sci. Instr. 40, 241 (1969); 42, 411 (1971)
8.10 M. Isaacson, D. Johnson: Ultramicroscopy 1, 33 (1975)
8.11 Y. Kihn, I. Sevely, B. Jouffrey: Philos. Mag. 33, 733 (1976)
8.12 C. Colliex, V.E. Cosslett, R.D. Leapman, P. Trebbia: Ultramicroscopy 1, 301 (1976)
8.13 J. Sevely, I.Ph. Perez, B. Jouffrey: In press
8.14 See the experimental work of Coster, Hanawalt, Lindsay and the theoretical interpretation of R. de L. Kronig: Z. Phys. 75, 468, 191 (1932)
8.15 I.I. Ritsko, S.E. Schnatterly, P.C. Gibbons: Phys. Rev. Lett. 32, 671 (1974)
8.16 T. Shiraiwa, T. Ishimura, M. Sawada: J. Phys. Soc. Jpn. 13, 847 (1958)
8.17 D.E. Saywers, E.A. Stern, F.W. Lytle: Phys. Rev. Lett. 27, 1204 (1971)
8.18 E.A. Stern: Phys. Rev. B 10, 3027 (1974)
8.19 B.M. Kincaid, A.E. Meixner, P.M. Platzman: Phys. Rev. Lett. 40, 1296 (1978)
8.20 G. Martens, P. Rabe, N. Schwenter and A. Werner, Phys. Rev. B 17, 1481 (1978)
8.21 P. Rabe, G. Tolkiehn, A. Werner, J. Phys. C: Solid State 12, 1173 (1979)
8.22 G.S. Brown, P. Eisenberger, P. Schmidt: Solid State Commun. 24, 201 (1977)
8.23 E.A. Stern, D.E. Sayers, J.G. Dash, H. Shechter, B. Bunker: Phys. Rev. Lett. 38, 767 (1977)
8.24 J.B. Boyce, T.M. Hayes, W. Stutius, J.C. Mikkelsen: Phys. Rev. Lett. 38, 1362 (1977)
9.1 E. Kröger: Z. Phys. 216, 115 (1968); 235, 403 (1970)
9.2 K. Zeppenfeld, H. Raether: Z. Phys. 193, 471 (1966)
9.3 H.R. Philipp, H. Ehrenreich: Phys. Rev. 129, 1550 (1963)
9.4 C. von Festenberg, E. Kröger: Phys. Lett. A 26, 339 (1968)
9.5 C. von Festenberg: Z. Phys. 214, 464 (1968); 227, 453 (1969)
9.6 C.H. Chen, J. Silcox, R. Vincent: Phys. Rev. B 12, 64 (1975)
9.7 D. Heitmann: Z. Phys. 245, 154 (1971)
9.8 C.H. Chen, J. Silcox, R. Vincent: Phys. Rev. Lett. 35, 390 (1975)
9.9 C.H. Chen, J. Silcox: Solid State Commun. 17, 273 (1975)
9.10 H.G. Wessjohann: Phys. Status Solidi (b) 77, 535 (1976)
9.11 J. Geiger, H. Katterwe: Solid State Commun. 16, 127 (1975); Optik 49, 25 (1977)
10.1 For further literature see:
10.1a K. Kliever, R. Fuchs: Adv. Chem. Phys. 27, 356 (1974)
10.1b E.N. Economu, K.L. Ngai: Adv. Chem. Phys. 27, 265 (1974)
10.1c H. Raether: Phys. Thin Films 7, 145 (1977)
10.2 R.H. Ritchie: Phys. Rev. 106, 874 (1957)
10.2a F. Wooten: *Optical Properties of Solids* (New York 1972)
10.3 E.A. Stern, R.A. Ferrell: Phys. Rev. 111, 1214 (1958)
10.3a R.H. Ritchie, H.B. Eldridge: Phys. Rev. 126, 1947 (1962)
10.4 R. Bruns, H. Raether: Z. Phys. 237, 98 (1970)
10.5 R.A. Ferrell: Phys. Rev. 111, 1214 (1958)

10.6 K.L. Kliewer, R. Fuchs: Phys. Rev. 153, 498 (1967)
10.7 E. Kretschmann, H. Raether: Z. Naturforsch. A 23, 2135 (1968)
10.8 D. Hornauer, H. Raether: Opt. Commun. 7, 297 (1973)
10.9 H. Kogelnik: "Integrated Optics", in *Topics in Applied Physics*, Vol. 7, ed. by T. Tamir (Springer, Berlin, Heidelberg, New York 1975) Chap. 2
10.10 D. Marcuse: *Theory of Dielectric Optical Waveguides* (Academic Press, 1979)
10.11 W. Steinmann: Phys. Status Solidi 28, 437 (1968)
10.12 R.H. Ritchie, H.B. Eldridge: Phys. Rev. 126, 1935 (1962)
10.13 J.C. Ashley: Phys. Rev. 155, 208 (1967)
10.14 E. Kröger: Z. Phys. 235, 403 (1970)
10.15 W. Steinmann: Phys. Rev. Lett. 5, 470 (1960); Z. Phys. 163, 92 (1961)
10.16 W.R. Brown, P. Wessel, E.P. Trounson: Phys. Rev. Lett. 5, 472 (1960)
10.17 H.D. Hattendorf: Phys. Status Solidi (a) 42, 489 (1977)
10.18 E.T. Arakawa, N.O. Davis, R.D. Birkhoff: Phys. Rev. A 135, 224 (1964)
10.19 R.J. Herickhoff, W.F. Hanson, E.T. Arakawa, R.D. Birkhoff: Phys. Rev. A 139, 1455 (1965)
10.20 A.J. Braundmeier, E.T. Arakawa: Opt. Commun. 2, 257 (1970)
10.21 R.J. Herickhoff, E.T. Arakawa, R.D. Birkhoff: Phys. Rev. A 137, 1433 (1965)
10.22 A.J. Braundmeier, E.T. Arakawa, M.W. Williams: Phys. Lett. A 32, 241 (1970)
10.23 B. Feuerbacher, B. Fitton: Phys. Rev. Lett. 24, 499 (1970)
10.24 H.O. Tittel: Phys. Lett. A 26, 145 (1969)
10.25 B. Schmalfeld: Unpublished work (1969)
10.26 A. Daudé, A. Savary, G. Jezequel, S. Robin: Opt. Commun. 1, 237 (1969)
10.27 D. Heitmann, V. Permien: Opt. Commun. 25, 196 (1978)
10.28 R.B. Pettit, J. Silcox, R. Vincent: Phys. Rev. B 11, 3116 (1975)
10.29 D. Schulz, M. Zurheide: Z. Phys. 211, 165 (1968)
10.30 S. Yamaguchi: J. Phys. Soc. Jpn. 17, 1172 (1962)
10.31 A.J. McAlister, E.A. Stern: Phys. Rev. 132, 1959 (1963)
10.32 M. Hattori, K. Yamada, H. Suzuki: J. Phys. Soc. Jpn. 18, 203 (1963)
10.33 J. Bösenberg: Z. Phys. 218, 282 (1969)
10.34 M. Skibowski, B. Feuerbacher, W. Steinmann: Z. Phys. 211, 329 (1968)
10.35 B. Feuerbacher: Thesis, Univ. of Munich (1968)
10.36 J. Brambring, H. Raether: Phys. Rev. Lett. 15, 882 (1965); Z. Phys. 199, 118 (1967)
10.37 P. Schreiber, H. Raether: Z. Naturforsch. A 21, 2116 (1966)
10.38 P. Schreiber: Z. Phys. 211, 257 (1968)
10.39 E.A. Stern: Phys. Rev. Lett. 19, 1321 (1967)
10.40 P. Pokrowsky, H. Raether: Surf. Sci. 83, 423 (1979)
10.41 H.G. Wessjohann: Thesis, Univ. Hamburg (1976); Phys. Status Solidi (b) 77, 535 (1976); Z. Phys. 269, 269 (1974)
10.42 H. Raether: Surf. Sci. 8, 233 (1967)
10.43 T. Kloos: Z. Phys. 208, 77 (1968)
10.43a T. Kloos: Z. Phys. 210, 303 (1968)
10.44 J. Krane: Unpublished work
10.45 J. Schilling: Z. Phys. B 25, 61 (1976)
10.46 M. Schlüter: Z. Phys. 247, 263 (1971); 250, 87 (1972);
10.46a A. A. Lucas, M. Sunjiec: Progress in Surf. Sci. 2, 2 (1972)
10.47 P. Schmüser: Z. Phys. 180, 105 (1964)
10.48 C. Kunz: Z. Phys. 167, 53 (1962)
10.49 J. Geiger: Z. Phys. 161, 243 (1961)
10.50 C. Kunz: Z. Phys. 196, 311 (1966)
10.51 K.J. Krane, H. Raether: Phys. Rev. Lett. 37, 1355 (1976)
10.52 J. Geiger: Phys. Status Solidi 24, 457 (1967)
10.53 A. Otto: Phys. Status Solidi 22, 401 (1967)
10.54 P. Zacharias: Z. Phys. 238, 172 (1970)
10.55 N.D. Lang: Solid State Commun. 7, 1047 (1969);
N.D. Lang, W. Kohn: Phys. Rev. B 1, 4555 (1970)
10.56 J. Langkowski: in H. Raether *Vac. Ultraviolet Radiat.*, Proc. Int. Conf., 4th, Hamburg (1974)

10.57 C.B. Duke, L. Pietronero, I.O. Porteus, I.F. Wendelken: Phys. Rev. B 12, 4059 (1975)
10.58 F. Forstmann, H. Stenschke: Phys. Rev. B 17, 1489 (1978)
10.58a A.D. Boardman, B.V. Paranjape, R. Teshima: Surf. Sci. 49, 275 (1975)
10.59 C.J. Powell, J.B. Swan: Phys. Rev. 115, 869 (1960)
10.60 J. Daniels: Z. Phys. 213, 227 (1968)
10.60a K.H. Gaukler: Z. Phys. 196, 85 (1966)
10.61 W.R. Miller, Jr., N.N. Axelrod: Solid State Commun. 3, 133 (1965)
10.62 M. Creuzburg: Z. Naturforsch. A 18, 101 (1963); Z. Phys. 174, 511 (1963)
10.63 C. Kunz, H. Raether: Solid State Commun. 1, 214 (1963)
10.64 C. Kunz: Z. Phys. 180, 127 (1964)
10.65 J. Daniels: Z. Phys. 203, 235 (1967)
10.66 M. Creuzburg, H. Raether: Z. Phys. 171, 436 (1963)
10.67 J. Lohff: Z. Phys. 171, 442 (1963)
10.68 C.J. Powell: Phys. Rev. 175, 972 (1968)
10.69 J. Lecante, Y. Ballu, D.M. Newns: Phys. Rev. Lett. 38, 36 (1977)
10.70 H. Froitzheim: "Electron Energy Loss Spectroscopy" in *Topics in Current Physics*, Vol. 4 (Springer, Berlin, Heidelberg, New York 1977) pp. 205-250
10.71 H. Ibach, J.E. Rowe: Phys. Rev. B 9, 1951 (1974)
10.72 J. Küppers: Surf. Sci. 36, 53 (1973)
10.73 G.W. Simmons, E.J. Scheibner: J. Appl. Phys. 43, 693 (1972)
10.74 H. Hopster, H. Ibach: Surf. Sci. 77, 109 (1978)
10.75 W. Erley, H. Wagner, H. Ibach: Surf. Sci. (in press)
10.76 M.R. Barnes, R.F. Willis: Phys. Rev. Lett. 41, 1729 (1978)
10.77 E. Evans, D.L. Mills: Phys. Rev. B 5, 4126 (1972); 7, 853 (1973)
10.78 D.M. Newns: Phys. Lett. A 60, 461 (1977)
10.79 Y.U. Romanow: Radio Phys. 7, 67 (1974)
10.80 P. Oelhafen: Surf. Sci. 47, 422 (1975)
10.81 P. von Blanckenhagen, H. Boersch, D. Fritsche, H.G. Seifert, G. Sauerbrey: Phys. Lett. 11, 4 (1964)
10.82 E.A. Stern: in *Optical and Electronic Structure*, ed. by F. Abeles (North Holland, Amsterdam 1966) p. 396
10.83 H. Boersch, P. Dobberstein, D. Fritzsche, G. Sauerbrey: Z. Phys. 187, 97 (1965)
10.84 P. Dobberstein, G. Sauerbrey: Phys. Lett. A 31, 6 (1970)
10.85 U. Bürker, W. Steinmann: Phys. Rev. Lett. 21, 3 (1968)
10.86 Y. Teng, E.A. Stern: Phys. Rev. Lett. 19, 511 (1967)
10.87 D. Heitmann: J. Phys. C: Solid State 10, 397 (1977); Phys. Status Solidi (b) 88, 493 (1978)
10.88 E. Kröger, H. Raether: Z. Phys. 244, 1 (1971)
10.89 I. Pockrand, H. Raether: Opt. Commun. 17, 353 (1976)
10.90 I. Pockrand: J. Phys. D: Appl. Phys. 9, 2423 (1976)
10.91 J.J. Cowan, E.P. Arakawa: Z. Phys. 235, 97 (1970)
10.92 W. Rothballer: Opt. Commun. 20, 429 (1977)
10.93 E.H. Rosengart, I. Pockrand: Opt. Lett. 1, 194 (1977)
10.94 M.C. Hutley, V.M. Bird: Opt. Acta 20, 771 (1973);
M.C. Hutley: Opt. Acta 20, 607 (1973);
J.E. Steward, W.S. Gallaway: Appl. Opt. 1, 421 (1962);
R.C. McPhedran, D. Maystre: Opt. Acta 21, 413 (1974)
10.95 E. Kretschmann: Z. Phys. 241, 313 (1971)
10.96 A. Otto: Phys. Status Solidi 26, K99 (1968); Z. Phys. 216, 398 (1968)
10.97 A. Otto: Z. Angew. Phys. 27, 207 (1969)
10.98 G. Borstel, H.J. Falge, A. Otto: in *Springer Tracts in Modern Physics*, Vol. 74 (Springer, Berlin, Heidelberg, New York 1974)
10.99 E. Kretschmann: Opt. Commun. 26, 41 (1978)
10.100 J. Bösenberg: Phys. Lett A 37, 439 (1971)
10.101 J. Bösenberg: Z. Phys. B 22, 267 (1975)
10.102 C. Maczek, A. Otto, W. Steinmann: Phys. Status Solidi B 51, K59 (1972)
10.103 I. Pockrand: Surf. Sci. 72, 577 (1978)

10.104 K. Holst, H. Raether: Opt. Commun. 2, 312 (1970)
10.105 J.G. Gordon, II, J.D. Swalen: Opt. Commun. 22, 374 (1977)
10.106 I. Pockrand, J.D. Swalen, J.G. Gordon, II, M.R. Philpott: Surf. Sci. 74, 237 (1978)
10.107 G. Wähling: Z. Naturforsch. A 33, 536 (1978)
10.108 H. Twietmeyer: Unpublished work (1975)
10.109 G. Schröder: In press
10.110 T. Lopez-Rios: Opt. Commun. 17, 342 (1976); T. Lopez-Rios, G. Vuye, II: Nuovo Cimento (b) 39, 823 (1977)
10.111 I. Pockrand, J.D. Swalen: J. Opt. Soc. Am. 68, 1147 (1978)
10.112 G. Wähling, D. Möbius, H. Raether: Z. Naturforsch. A 33, 907 (1978)
10.113 I. Pockrand, J.D. Swalen, R. Santo, A. Brillante, M.R. Philpott: J. Chem. Phys. 69, 4001 (1978)
10.113a R. Kötz, D.M. Kolb, J.K. Sass: Surf. Sci. 69, 359 (1977)
10.114 E.A. Stern: Phys. Rev. Lett. 19, 1321 (1967)
10.115 E. Kretschmann, H. Raether: Z. Naturforsch. A 22, 615 (1967)
10.116 E. Kretschmann, E. Kröger: J. Opt. Soc. Am. 65, 150 (1975)
10.117 E. Kröger, E. Kretschmann: Phys. Status Solidi 76, 515 (1976)
10.118 E. Kretschmann, Opt. Commun. 10, 353 (1971)
10.119 E. Kretschmann, Opt. Commun. 5, 331 (1972)
10.120 E. Kretschmann, Opt. Commun. 6. 185 (1972)
10.121 J. Bodesheim, A. Otto: Surf. Sci. 45, 441 (1974)
10.122 D. Hornauer: Opt. Commun. 16, 76 (1976)
10.123 D. Hornauer, H. Kapitza, H. Raether: J. Phys. D 7, L100 (1974)
10.124 A.J. Braundmeier, E.T. Arakawa: J. Phys. Chem. Solids 35, 517 (1974)
10.125 R. Orlowski, H. Raether: Surf. Sci. 54, 303 (1976)
10.126 H. Kapitza: Opt. Commun. 16, 73 (1976)
10.126a R. Orlowski, P. Urner, D. Hornauer: Surf. Sci. 82, 69 (1979)
10.127 F. Toigo, A. Marvin, V. Celli, N.R. Hill: Phys. Rev. B 15, 5618 (1977); further literature to this problem see A.A. Maradudin, D.L. Mills: Phys. Rev. B 11, 1392 (1975) and D.L. Mills: Phys. Rev. B 12, 4036 (1975)
10.128 I. Pockrand, H. Raether: Opt. Commun. 18, 395 (1976)
10.129 I. Pockrand: Opt. Commun. 13, 311 (1975)
10.130 I. Pockrand: J. Phys. D 9, 2423 (1976)
10.131 I. Pockrand, H. Raether: Appl. Opt. 16, 1784 (1977)
10.132 E.H. Rosengart: Unpublished work (1978)
10.133 R. Fuchs, K.L. Kliewer: Phys. Rev. A 140, 2076 (1965); K.L. Kliewer, R. Fuchs: Phys. Rev. 144, 495 (1966)
10.134 H. Boersch, J. Geiger, W. Stickel: Phys. Rev. Lett. 17, 379 (1966); Z. Phys. 212, 130 (1968)
10.135 H. Ibach: Phys. Rev. Lett. 24, 1416 (1970)
10.136 N. Marschall, B. Fischer: Phys. Rev. Lett. 28, 811 (1972)
10.136a J. Lagois, B. Fischer: Phys. Rev. Lett. 36, 680 (1976); Phys. Rev. B 17, 3814 (1978); *Topics in Current Physics*, Vol. 14 (Springer 1969) p. 183
10.137 H. Boersch, J. Geiger, A. Bohg: Z. Phys. 227, 141 (1969); J. Geiger, R.H. Jakobs: Z. Phys. 271, 323 (1974)
10.138 I. Hirabayashi, T. Koda, Y. Tokura, J. Murata, Y. Kaneko: J. Phys. Soc. Jpn. 40, 1215 (1976)
10.139 M.G. Sceats, K. Tomioka, St.A. Rice: J. Chem. Phys. 66, 4486 (1977)
10.140 M.R. Philpott, A. Brillante, I. Pockrand, J.D. Swalen: Mol. Cryst. Liq. Cryst. (in press)
10.141 I. Pockrand, A. Brillante, M.R. Philpott, J.D. Swalen: Opt. Commun. 27, 91 (1978)
10.142 F. Fujimoto, K. Komaki: J. Phys. Soc. Jpn. 25, 1679 (1968)
10.143 H. Petersen: Solid State Commun. 23, 931 (1977)
10.144 U. Kreibig, P. Zacharias: Z. Phys. 231, 128 (1970)
10.145 J.C. Ashley, L.C. Emerson: Surf. Sci. 41, 615 (1974)
10.146 C.A. Pfeiffer, E.N. Economou, K.L. Ngai: Phys. Rev. B 10, 3038 (1974)
10.147 C. Miziumski: Phys. Lett. A 40, 187 (1972)
10.148 F. Stern: Phys. Rev. Lett. 18, 546 (1967)
10.149 W.T. Sommer, T.J. Tanner: Phys. Rev. Lett. 27, 1345 (1971)

10.150 C.C. Grimes, G. Adams: Phys. Rev. Lett. 36, 145 (1976)
10.151 R.S. Crandall, E. Williams: Phys. Lett. A 34, 404 (1971)
10.152 C.C. Grimes, G. Adams: Phys. Rev. Lett. 42, 795 (1979)
10.153 S.J. Allen, Jr., D.C. Tsue, R.A. Logan: Phys. Rev. Lett. 38, 980 (1977)
10.154 T.N. Theis, J.O. Kotthaus, P.J. Stiles: Solid State Commun. 24, 273 (1977)
10.155 A.V. Chaplik: Sov. Phys. JETP 35, 395 (1972)
10.156 H.R. Zeller: *Advances in Solid State Physics XIII* (Vieweg 1973) p. 31
10.157 J.J. Ritsko, D.J. Sandmann, A.J. Epstein , P.C. Gibbons, S.E. Schnatterly, J. Fields: Phys. Rev. Lett. 34, 1330 (1975)
10.158 C.H. Chen, I. Silcox, A.F. Garito, A.I. Heeger, A.G. McDianuid: Phys. Rev. Lett. 36, 525 (1976)
10.159 P.F. Williams, A.N. Bloch: Phys. Rev. B 10, 1097 (1974)
10.160 A. Adams, J.C. Wyss, P.K. Hansma: Phys. Rev. Lett. 42, 912 (1979)
10.161 P.K. Hansma, H.P. Broida: Appl. Phys. Lett. 32, 545 (1978)
10.162 R.W. Rendell, D.J. Scalpino, B. Mühlschlegel: Phys. Rev. Lett. 41, 1746 (1978)
10.163 H. Froitzheim, H. Ibach, D.L. Mills: Phys. Rev. B. 11, 4980 (1975)
10.164 G. Chiarotti, S. Nannarone, R. Pastore, P. Chiaradia: Phys. Rev. B 4, 3398 (1971)
11.1 R. Speidel, D. Kurz: Optik 49, 173 (1977)
11.2 M. Isaacson, R. Gomer: Appl. Phys. 15, 253 (1978)
11.3 K.H. Gaukler, R. Speidel, F. Vorster: Optik 42, 391 (1975)
11.4 R.B. Pettit, J. Silcox, R. Vincent: Phys. Rev. B 11, 3116 (1975)
11.5 P. Keil: Z. Naturforsch. A 21, 503 (1966)
11.6 W. Legler: Z. Phys. 171, 424 (1963)
11.7 W.A.M. Hartl: Z. Phys. 191, 487 (1966)
11.8 U. Killat: J. Phys. C: Solid State Phys. 7, 2396 (1974)
11.9 K.D. Sevier: *Low Energy Electron Spectroscopy* (Wiley 1972)
11.10 K. Brack: Z. Naturforsch. A 17, 1066 (1962)
11.11 W. Glaser: *Elektronenoptik* (Springer, Wien 1952)
11.12 H. Boersch, J. Geiger, W. Stickel: Z. Phys. 180, 415 (1964)
11.13 W.H.J. Andersen, J.B. Le Poole: J. Phys. E 3, 121 (1970)
11.14 G.H. Curtis, J. Silcox: Rev. Sci. Instr. 42, 630 (1971)
11.15 T. Takigawa, F. Fujimoto: J. Phys. Soc. Jpn. 40, 1630 (1976)
11.16 P.C. Gibbons, J.J. Ritsko, S.E. Schnatterly: Rev. Sci. Instr. 46, 1546 (1975)
11.16a J. Lohff: Z. Phys. 171, 442 (1963)
11.17 T. Ichinokawa: Jpn. J. Appl. Phys. 7, 799 (1968)
11.18 J. Sevely, J.P. Perez, B. Jouffrey: C.R. Acad. Sc. (Paris) 276, 515 (1973)
11.19 G. Zanchi, I.Ph. Perez, I. Sevely: Optik 43, 495 (1975)
11.20 G. Zanchi, I. Sevely, B. Jouffrey: J. Microsc. Spectrosc. Electron. 1, 95 (1977)
11.21 H. Froitzheim, H. Ibach: Z. Phys. 269, 17 (1974)
11.22 J. Schilling: Z. Phys. B 25, 61 (1976)

Subject Index

Springer Series in Solid-State Sciences

Editors: M. Cardona, P. Fulde, H.-J. Queisser

Volume 1
C. P. Slichter
Principles of Magnetic Resonance
2nd revised and expanded edition. 1978.
ISBN 3-540-08476-2
(Originally published by Harper & Row Publ., New York)

Volume 2
O. Madelung
Introduction to Solid-State Theory
Translated from the German by B. C. Taylor
1978.
ISBN 3-540-08516-5

Volume 3
Z. G. Pinsker
Dynamical Scattering of X-Rays in Crystals
1978.
ISBN 3-540-08564-5

Volume 4
Inelastic Electron Tunneling Spectroscopy
Proceedings of the International Conference, and Symposium on Electron Tunneling, University of Missouri-Columbia, USA, May 25–27, 1977
Editor: T. Wolfram
1978.
ISBN 3-540-08691-9

Volume 5
F. Rosenberger
Fundamentals of Crystal Growth I
Macroscopic Equilibrium and Transport Concepts
1979.
ISBN 3-540-09023-1

Volume 6
R. P. Huebener
Magnetic Flux Structures in Superconductors
1979.
ISBN 3-540-09213-7

Springer-Verlag
Berlin Heidelberg NewYork

Volume 7
E. N. Economou
Green's Functions in Quantum Physics
1979.
ISBN 3-540-09154-8

Volume 8
Solitons and Condensed Matter Physics
Proceedings of the Symposium on Nonlinear (Soliton) Structure and Dynamics in Condensed Matter Oxford, England, June 27–29, 1978
Editors: A. R. Bishop, T. Schneider
1978.
ISBN 3-540-09138-6

Volume 9
V. M. Fridkin
Photoferroelectrics
1979.
ISBN 3-540-09418-0

Volume 10
H. Bilz, W. Kress
Phonon Dispersion Relations in Insulators
1979.
ISBN 3-540-09399-0

Volume 11
B. R. Nag
Electron Transport in Compound Semiconductors
1980.
ISBN 3-540-09845-3

Volume 13
The Physics of Selenium and Tellurium
Proceedings of the International Conference On the Physics of Selenium and Tellurium Königstein, Fed. Rep. of Germany, May 28–31, 1979
1979.
ISBN 3-540-09692-2

Volume 14
A. H. Eschenfelder
Magnetic Bubbles
1980.
ISBN 3-540-09822-4

Excitons

Editors: K. Cho
1979. 118 figures, 8 tables. XI, 274 pages
(Topics in Current Physics, Volume 14)
ISBN 3-540-09567-5

Contents:
K. Cho: Introduction. – *K. Cho:* Internal Structure of Excitons. – *P. J. Dean, D. C. Herbert:* Bound Excitons in Semiconductors. – *B. Fischer, J. Lagois:* Surface Excitons Polaritons. – *P. Y. Yu:* Study of Excitons and Exciton-Phonon Interactions by Resonant Raman and Brillouin Spectroscopies.

Light Scattering in Solids

Editor: M. Cardona
1975. 111 figures, 3 tables. XIII, 339 pages
(Topics in Applied Physics, Volume 8)
ISBN 3-540-07354-X

Contents:
M. Cardona: Introduction. – *A. Pinczuk, E. Burstein:* Fundamentals of Inelastic Light Scattering in Semiconductors and Insulators. – *R. M. Martin, L. M. Falicov:* Resonant Raman Scattering. – *M. V. Klein:* Electronic Raman Scattering. – *M. H. Brodsky:* Raman Scattering in Amorphous Semiconductors. – *A. S. Pine:* Brillouin Scattering in Semiconductors. – *Y.-R. Shen:* Stimulated Raman Scattering.

Photoemission in Solids I

General Principles

Editors: M. Cardona, L. Ley
1978. 90 figures, 17 tables. XI, 290 pages
(Topics in Applied Physics, Volume 26)
ISBN 3-540-08685-4

Contents:
M. Cardona, L. Ley: Introduction. – *W. L. Schaich:* Theory of Photoemission: Independent Particle Model. – *S. T. Manson:* The Calculation of Photoionization Cross Sections: An Antomic View. – *D. A. Shirley:* Many-Electron and Final State Effects: Beyond the One-Electron Picture. – *G. K. Wertheim, P. H. Citrin:* Fermi Surface Excitations in X-Ray Photoemission Line Shapes From Metals. – *N. V. Smith:* Angular Dependent Photoemission. – Appendix.

Photoemission in Solids II

Case Studies

Editors: L. Ley, M. Cardona
1979. 214 figures, 26 tables. XVIII, 401 pages
(Topics in Applied Physics, Volume 27)
ISBN 3-540-09202-1

Contents:
L. Ley, M. Cardona: Introduction. – *L. Ley, M. Cardona, R. A. Pollak:* Photoemission in Semiconductors. – *S. Hüfner:* Unfilled Inner Shells: Transition Metals and Compounds. – *M. Campagna, G. K. Wertheim, Y. Baer:* Unfilled Inner Shells: Rare Earths and Their Compounds. – *W. D. Grobman, E. E. Koch:* Photoemission from Organic Molecular Crystals. – *C. Kunz:* Synchrotron Radiation: Overview. – *P. Steiner, H. Höchst, S. Hüfner:* Simple Metals. – Appendix: Table of Core. – Level Binding Energies. – Additional References with Titles. – Subject Index.

Springer-Verlag
Berlin
Heidelberg
New York

Made in the USA
Monee, IL
07 May 2026

49715171R00116